A Manual of Forensic Entomology

A Manual of Forensic Entomology

Kenneth G. V. Smith
Department of Entomology, British Museum
(Natural History)

The Trustees of the British Museum (Natural History)

London 1986

© Trustees of the British Museum (Natural History) 1986

All rights reserved. Except for brief quotations in a review,
this book, or parts thereof, must not be reproduced in any
form without permission in writing from the publisher.
British Museum (Natural History), Cromwell Road, London SW7 5BD.

First published 1986 by the British Museum (Natural History) and
Cornell University Press

British Library Cataloging in Publication Data
Smith, Kenneth G. V.
   A manual of forensic entomology.
   1. Forensic entomology
   I. Title.
   599.7'0024614   RA1063.45

ISBN 0-565-00990-7

Printed at the University Printing House, Oxford

Dedicated to

JEAN PIERRE MÉGNIN
(1828–1905)

MARCEL LECLERCQ and PEKKA NUORTEVA

Pioneers in the application of entomology to forensic science

'Succession is no doubt one of the most important and widespread of the phenomena discovered by ecologists up to the present time.'

V. E. Shelford, 1911

HAMLET: How long will a man lie i' th' earth ere he rot?
FIRST CLOWN: ... A tanner will last you nine year
HAMLET: Why he, more than another?
FIRST CLOWN: Why sir, his hide is so tann'd with his trade, that he will keep out water a great while. And your water, is a sore decayer of your whoreson dead body. . . .

Hamlet, Act V, Scene 1

Contents

Acknowledgements	9
Introduction	11
The Faunal Succession on cadavers	13
Exposed corpses	13
Buried corpses	17
Corpses immersed in water	25
Mummification	27
Burnt bodies	27
Geographical location	28
Temperature and humidity	30
Light and shade	31
Seasonal and daily periodicity	33
Availability of food and competition	34
Manner of death	34
Methods and Techniques	37
Equipment for collection and preservation	37
Procedure on site	39
Procedure in the laboratory	41
Interpretation of data for forensic use	47
Experimental work with carrion	50
Life histories	52
Identification	53
Case histories	56
Diptera (Flies and maggots)	68
Identification key to adult flies visiting carrion	68
Identification key to maggots on carrion	72
Nematocera	73
Trichoceridae	73
Psychodidae	75
Brachycera	77
Stratiomyidae	77
Cyclorrhapha – Aschiza	77
Phoridae	77
Syrphidae	80
Acalyptratae	81
Dryomyzidae	81
Coelopidae	81
Heleomyzidae	81
Sepsidae	84
Sphaeroceridae	87
Piophilidae	90
Ephydridae	95
Drosophilidae	96
Milichiidae	99
Calyptratae	99
Sarcophagidae	99
Calliphoridae	102
Fanniidae	120
Muscidae	122

Coleoptera (Beetles and larvae) 138
Lepidoptera (Butterflies, moths and caterpillars) 150
Hymenoptera (Bees, wasps, ants, etc.) 154
Dictyoptera, Orthoptera, etc. (Cockroaches, crickets, grasshoppers) 157
Hemiptera (Plant-bugs, aphids, etc.) 158
Collembola (Springtails) 160
Isoptera (Termites) 161
Ectoparasites (Siphonaptera (fleas) and Phthiraptera (lice)) 162
Arthropods other than insects (Millipedes, spiders, mites, etc.) 165
Insects and transport 167
Cannabis insects 169
Glossary 174
Bibliography 178
Index 195

Acknowledgements

The following are warmly thanked for very kindly allowing me to use illustrations or information from their published or unpublished works:

Professor David Bellamy, University of Durham, for use of his photograph of the musk ox remains (Fig. 14).

Dr George F. Bornemissza for the gift of a long sought after copy of his classic paper in the *Australian Journal of Zoology* and for his permission, and the permission of the CSIRO Editorial and Publications Service, Melbourne, to include two of his figures (Figs 3, 5).

Mr Alan Brindle for permission to base some drawings on his published illustrations.

Dr B. W. Cornaby and the Editor of *Biotropica* for permission to use his food-web diagram (Fig. 7).

The Superintendent, Devon and Cornwall Constabulary, for permission to quote some case details (Case history 19, p. 66).

Dr Paul Freeman and the Royal Entomological Society of London for use of his figure of *Bradysia* from his Sciaridae volume in the Society's *Handbooks for the Identification of British Insects* series (Fig. 394).

Dr Ammon Freidberg and the Editor of *Entomologica Scandinavica* for permission to reproduce his figures of *Centrophlebomyia furcata* (Figs 150–153) and other information on 'Thyreophoridae'.

Dr Norman Hickin for permission to reproduce his scraper board drawing of a caddis-fly larva (Fig. 380).

Dr A. K. Kamal and the Entomological Society of America for permission to reproduce a table from *Annals of the Entomological Society of America*, **51**, (1958) (Table 5).

Professor Kiyotoshi Kaneko, Aichi Medical University, for permission to base two figures on his published illustrations of Phoridae larvae (Figs 76, 77).

Dr Marcel Leclercq, Université de Liège, and Pergamon Press for permission to quote a case history and other information from his book *Entomological Parasitology* (1969).

Dr H. Lundt and VEB Gustav Fischer Verlag, Jena, for permission to reproduce his figure 4 from *Pedobiologia* **4** (1964; Fig. 8).

Dr J. F. McAlpine, Biosystematics Research Institute, Ottawa, for use of his photograph of the lemming carcase (Fig. 13), and Mrs Nancy Ellacott for use of the figure of *Trichocera garretti* Alexander from McAlpine *et al.* (1981), *Manual of Nearctic Diptera*, reproduced here by permission of the Minister of Supply and Services, Canada (Fig. 291).

Dr L. Nabaglo, Instutut Ekologii PAN, Warsaw for permission to reproduce illustrations from his work in *Ekologia Polska* (1973) (Figs 9, 17, 18).

Dr Sumio Nagasawa of Kyoto University, and Dr Toshiaki Ikeshoji, Editor of the *Japanese Journal of Applied Entomology and Zoology*, for permission to reproduce the graph illustrating Pradhan's formula from Nagasawa & Kishino (1965) (Fig. 15).

Dr G. E. J. Nixon, Mr A. Smith and Hutchinson, London are thanked for use of Figs 367–8.

Professor Pekka Nuorteva, University of Helsinki, for generous permission to use illustrations and to quote freely from his important publications on forensic entomology in *Annales Entomologici Fennici*, *Forensic Science* and his chapter in *Forensic Medicine, a Study in Trauma and Environmental Hazards*, Saunders, 1977 (Figs 4, 32, 47, 48).

Professor Emeritus Toyohi Okada, Metropolitan University of Tokyo, for the generous gift of his important book *Systematic Study of the Early Stages of Drosophilidae* (1968) on which some of my figures of *Drosophila* are based.

Professor Jerry A. Payne, his associates and publishers for permission to use figures from important papers in *Ecology*, *Nature* and *Journal of the Georgia Entomological Society* (Figs 1, 2, 10, 11, 12, 19).

Dr H. B. Reed and Professor Robert P. McIntosh for permission to reproduce a diagram from Dr Reed's paper in the *American Midland Naturalist* (1958; Fig. 6).

Dr Christian Reiter, Institute for Legal Medicine of the University of Vienna, and the publishers Springer Verlag of Heidelberg, for permission to reproduce his important 'Isomegalendiagram' (*Zeitschrift für Rechtsmedizin* **91**, 1984) and other information (Figs 16, 45).

The Librarian of the Royal Society of Medicine for access to certain journals otherwise not available to me.

Dr H. Schumann, Humboldt University, Berlin, for generous permission to base several of my drawings on his many fine published illustrations of cyclorrhaphous Diptera larvae.

Professor Eugène Séguy, Paris, for permission to use figures from his works (*Muscina*, 1923a and *Centrophlebomyia furcata*, 1950) (Figs 280, 281, 301).

Dr Monique S. J. Simmonds, Birkbeck College, London, for a sight of her important thesis on *Parasitoids of Synanthropic Flies* (1984).

Professor Keith Simpson and his publishers, Edward Arnold Ltd, London, for permission to quote information from his classic book *Forensic Medicine*, now in its ninth edition (1985).

Professor J. A. Slater and Dr R. M. Baronowski, and their publishers, Wm C. Brown, Dubuque, for permission to use illustrations from their book *How to Know the True Bugs* (1978; Figs 374–379).

Dr Ellen Thomsen for permission to use illustrations from her late husband's work (Thomsen, M., 1934; Fig. 31).

The United States Department of Agriculture for permission to copy illustrations from Laake, Cushing & Parish's 1936 paper on *Cochliomyia* and James' 1947 paper on myiasis (Figs 234–240, 88 & 89, respectively).

Mr Geoff Willott and his colleagues at the Metropolitan Police Forensic Science Laboratories for involvement in many of their interesting cases.

The Zoological Institute of the USSR Academy of Sciences, Leningrad ('Nauka' publishers) and VAAP (Soviet Copyright Agency) for permission to reproduce some excellent figures of Diptera which appeared in the late Professor A. A. Stakelberg's book [*Diptera associated with man from the Russian fauna*] (1956; Figs 292, 295–299, 303–311, 313–315.)

Dr Fritz Zumpt, Johannesburg, and Butterworths, London, for permission to include some information and illustrations from his classic book *Myiasis in Man and Animals in the Old World* (1965).

I thank Peter York and Tony Gowing, Photographic Section, British Museum (Natural History), for photographing microscopical preparations (Fig. 44), and several book illustrations, respectively.

I thank the following colleagues and friends who have carefully read sections of the book and offered valuable critical comments, though they are not of course responsible for any errors remaining: Dr L. A. Mound, Richard I. Vane-Wright, Adrian C. Pont, Brian Pitkin, David Carter, Bill Dolling and Dr Jane Marshall.

I thank the following colleagues and friends for help with information and advice: James Dear, Dr Alan M. Easton, Dr Zak Erzinçlioğlu, Dr Bernard Greenberg, Dr W. T. Hendry, Paul D. Hillyard, Keith Hyatt, Dr Richard P. Lane, Dr Wayne D. Lord, Dr Leif Lyneborg, Dr W. A. Sands, Dr Milton W. Shaw, Dr John Smart, Dr F. Christian Thompson, Dr Charles Vincent and Fred R. Wanless.

Finally I thank my wife for her careful and critical preparation of the typescript, her enthusiastic support, sympathetic understanding and tolerance during the course of this frequently unsavoury project, and for cheerfully coping with all those domestic responsibilities I so readily shirked while writing the book.

Kenneth G. V. Smith
1985

Introduction

The vast majority of insects have very little direct association with man: they are generally biologically beneficial as pollinators of plants and as basic members of important food chains supporting higher forms of life. Man, through his exploding populations with their relentless colonization of the planet and modification of the natural environment, has been brought into contact or even conflict with insects in various ways. They can eat his crops, parasitize and spread disease among his domesticated animals, and finally attack his person either as blood-sucking external parasites, when they may incidentally transmit diseases, or by invading his tissues as facultative or obligate parasites. The principal form of tissue invasion is by the larvae of flies (Diptera) and this condition is called myiasis.

After death the tissue of animals, including man, are still attractive to a variety of insects and other invertebrates. Not surprisingly, flies, especially their larvae or maggots, figure largely in this fauna and include some of the species involved in myiasis.

Until the Seventeenth Century it was believed that the presence of 'worms' (maggots) in corpses was due to spontaneous generation. In 1668 Francisco Redi (see Guiart, 1898) proved by a series of experiments that these larvae came from the eggs of flies deposited on the putrefying carcases, and he distinguished four species. Linnaeus (1767) had already observed that 'three flies consume the corpse of a horse as quickly as a lion did' (Leclercq, 1969). Réamur (1738) and Macquart (1835) were also impressed by the fecundity of blowflies.

The insects and other invertebrates feeding on carrion form a distinct faunal succession associated with the various stages of decay. Recognition of the species involved in their different immature stages in the succession, coupled with a knowledge of their rates of development, can thus give an indication of the age of the corpse. Because of this, entomologists may be called upon to identify specimens for medico-legal purposes, particularly as an aid in establishing the time of death, an all important factor in murder cases.

The earliest application of entomology in forensic science usually quoted is by Bergeret (see Yovanovitch, 1888 and Case histories). However, Keh (1985) draws attention to a thirteenth-century Chinese manual on forensic medicine (see McKnight, 1981) in which the following entomological case is cited. Following a murder by sickle a number of farmers were assembled with their sickles laid before them upon the ground. Flies settled on only one sickle, the owner of which then confessed. The classic work in the field, which also enabled some identifications to be made, was by Mégnin (1887, 1894). Since that time entomological evidence has been used sporadically in several murder cases in Britain and more frequently on the Continent with increasing success. Summaries have been provided by the outstanding exponents of the technique, Leclercq (1969, 1978) and Nuorteva (1977). However, the only forensically orientated works providing information on the *identification* of *some* cadaver fauna since Mégnin (1894) are the brief accounts by Easton & Smith (1970), Lord & Burger (1983), Reiter (1984), and Reiter & Wollenek, (1982, 1983a, b).

The main aim of the present work is to rectify this situation, principally for the British and European fauna, but also for some of the more important non-European genera.

One of the main handicaps to the development of research in this field has been the limitation of investigations to animal carrion. For obvious reasons, largely ethical and religious, human corpses have not been readily available for the detailed experimental field studies ideally required. Nevertheless, much valuable information has come from early studies of human cadavers and faunistic work on animal carrion – which has long been recognized as a fruitful source of specimens by entomologists. The published work in this field is fully reviewed in the text and included in the bibliography on a world basis in the present work. All published references to records from human corpses are also included and briefly commented upon. Techniques for the application of entomology to forensic investigations are fully described. Some cases illustrating the successful application of entomology in forensic work are cited in full and suggestions are made for fruitful areas of future research.

The more theoretical aspects of carrion as a functioning ecological system are beyond the scope of this book and the excellent introduction by Putman (1983) should be consulted.

While the study of the cadaver fauna constitutes the most important forensic application of entomology, there are other medico-legal applications involving sudden death, traffic accidents, criminology and occupational diseases. These are discussed, where appropriate, in the taxonomic sections of the text and, where of sufficient importance, are treated in separate sections towards the end of the book e.g. Cannabis Insects, Insects and Transport.

An elementary knowledge of entomology is assumed, but as far as possible technical terms have been kept to a minimum and scientific names are supplemented with vernacular names where these are available. A glossary of technical terms used in the book, or likely to be encountered in following up the scientific literature referred to, is given at the end of the book. Short introductions to the subject not cited above and of use to students without English as a first language are provided by: Leclercq, 1968 (Flemish); Leclercq, 1975, 1978 (French); Hennig, 1950 (German); de Stefani, 1921 (Italian); Kano, 1966 (Japanese); Aruzhonov, 1963, Lopatenok *et al.*, 1964, Marchenko, 1978, Rubezhanskiv, 1965, Rubezhanskiv & Ozanovskii 1964 (Russian); and López & Gisbert, 1962 and Vargas, 1977 (Spanish).

A bibliography of forensic entomology (Vincent *et al.*, 1985) has appeared as this book goes to press. It includes many more references to earlier literature than are given here, but fewer identification sources. It is hoped that the two works together will provide a useful basis for the future development of the subject.

Anyone considering entering the forensic field would do well to contemplate carefully the non-entomological aspects of the work and the responsibility involved (see Meek *et al.*, 1983; Smith & Dear, 1978; Wilson, 1982).

The faunal succession on cadavers

There are various methods of establishing time of death for a human corpse, e.g. histological, chemical, bacteriological and zoological. The last method is based on a faunistic study of the cadaver and can give accurate results if the data are carefully collected.

The invertebrate fauna of carrion consists mostly of insects and in its simplest form the forensic application is based on a study of sequence with which sarcosaprophagous insects appear on the cadaver. The recognition of each species in all stages and a knowledge of the time occupied by each stage at varying temperatures enables an estimate of time of death to be made. It may also be possible to establish if a body has been moved or partly concealed during decomposition.

Not all of the invertebrates occurring on carrion are actually feeding on it and four ecological categories can be recognized in the carrion community:

1. Necrophagous species — Feed on the carrion itself and constitute the most important category in establishing time of death, e.g. Diptera: Calliphoridae (blowflies); Coleoptera (beetles): Silphidae (in part), Dermestidae.
2. Predators and parasites on the necrophagous species — Second most important forensic category, e.g. Coleoptera: Silphidae (in part), Staphylinidae; Diptera: some carrion feeders become predaceous in later instars, e.g. *Chrysomya* (Calliphoridae), *Ophyra* and *Hydrotaea* (Muscidae).
3. Omnivorous species — Wasps, ants and some Coleoptera feed both on the corpse and its inhabitants.
4. Adventive species — Use the corpse as an extension of their environment, e.g. Collembola (springtails), spiders (which may become incidental predators).

Other species may of course occur accidentally by seeking moisture or merely a resting place. Yet other species may drop or be knocked onto a corpse from surrounding vegetation, especially if an attempt has been made to conceal the corpse. The non-carrion associated members of the existing soil fauna soon vacate the site of a decomposing corpse and do not return for some time (Bornemissza, 1957).

Exposed corpses

Clearly the faunal succession on carrion is linked to the natural changes which take place in a corpse following death. These may be summarized for a human corpse as follows (adapted from Simpson, 1985 and Leclercq, 1969).

After death the body temperature falls to that of its surroundings. Ectoparasites then leave the body. The first major change is a stiffening of the muscle fibres related to the breakdown of glycogen and the accumulation of lactic acid. This takes some five to seven hours, spreading from the face downwards, and lasts some 48 to 72 hours. The duration of *rigor mortis* depends on the metabolic state at death; it may not develop fully at summer temperatures or may be prolonged by low temperatures. It develops earlier and disappears more quickly in babies.

There follows a series of biochemical fermentation processes (autolysis) resulting in the release of gases such as ammonia (NH_3), hydrogen sulphide (H_2S), carbon dioxide (CO_2) and nitrogen (N_2). During this stage the flanks become stained greenish and the body becomes bloated. Putrefaction then follows and is largely due to the action of micro-

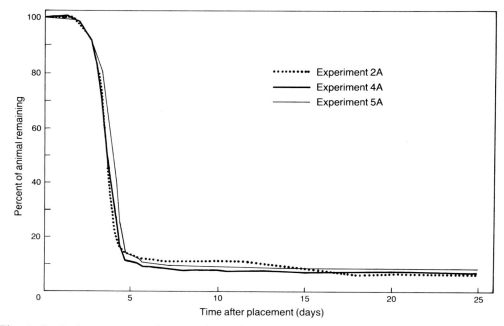

Figs 1, 2 1, decay curves showing loss of weight of pig carrion when exposed to insects during the summer of 1962. 2, decay curves showing loss of weight of pig carrion during the same period but *not* exposed to insects (after Payne, 1965).

---

organisms within the body, especially the intestinal flora, and later invading saprophytic fungi and bacteria. Bianchini (1930) recognized three waves of fungoid species following autolysis and up to the breakdown of the skeleton.

The necrophagous insects appear after the onset of autolysis and putrefaction, depending on the time of year and the situation of the corpse. Their activities accelerate putrefaction and the disintegration of the corpse (Figs 1, 2). Hobson (1932) found that initially larvae feed on liquid between the muscle fibres because the tissues are too acid. Later, when the tissues become alkaline, the intermuscular tissue is attacked.

The various parts of the body decompose at different rates and the usual order is:

1. Intestines, stomach, liver blood, heart blood and circulation, heart muscle
2. Air passages, lungs
3. Brain
4. Kidneys and bladder
5. Voluntary muscles
6. Uterus

This has a relevance to entomologists' investigations since some of the experimental field work on carrion fauna has been done on pieces of meat or liver and, clearly, the resultant data need careful interpolation when applied to a consideration of the situation in complete corpses. In slaughterhouses, blowflies are particularly attracted to entrails, cut surfaces and the kidney region on the carcase. Liver is said to be unattractive for three hours after removal from the carcase, but thereafter soon becomes infested (A. A. Green, 1951; and see Lindquist, 1954).

The number of waves of insects in the succession on carrion has been differently

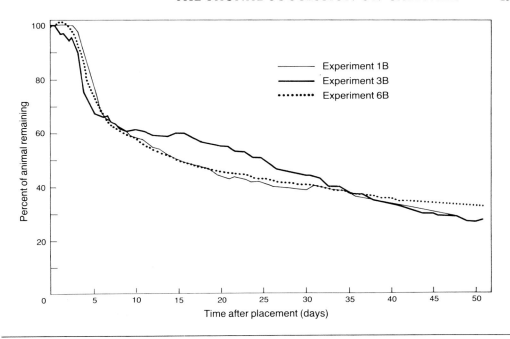

interpreted by different workers. It is not surprising that the lines of division may be indistinct, since biological systems are rarely that simple. Nevertheless, the existence of a succession is recognizable.

Mégnin (1894), in his classic work, recognized eight invasion waves of arthropods on dead human bodies. These waves are shown in Table 1 as tabulated by Johnston & Villeneuve (1897). Subsequent work has been based mostly on data from animal corpses, but de Stefani (1921) recognized six waves from a number of cases involving human cadavers in which he was able to ascertain time and place of death. In a study of sheep and cat carcases, Fuller (1934) recognized three stages, the last of which she stated to include several of Mégnin's later stages.

Howden (1950), quoted by Reed (1958), working with assorted carrion of partial and entire carcases in North Carolina recognized only two waves. Reed himself, working with dog carcases in Tennessee, recognized four waves as did Jirón & Cartín (1981) on dog in Costa Rica, Johnson (1975) with small mammals in Illinois and Rodriguez & Bass (1983) with human corpses in Tennessee. Utsumi (1958), working with dead dogs and rats in Japan, recognized only two waves.

These reductions in the number of waves were made in an attempt to define biological communities (biocenoses), but this complicates and lessens forensic applicability. As Nuorteva (1977) has pointed out, the forensic pathologist does not need the help of entomology to decide what *state* of decay a body has reached.

Payne (1965), in his impressive work on pig carcases in the USA, recognized six stages of decay: fresh, bloated, active, advanced, dry and remains. For each stage he defined the associated insect community and analysed the percentage abundance of species attracted to the various stages (Table 2). Lord & Burger (1984), working with seal carcasses on the New England Coast, recognised five stages of decomposition.

Bornemissza (1957), working with dead guinea pigs in Australia, made a most important study of the carrion community. He recognized five stages of carcase decomposition as follows:

Table 1 Principal members of the faunal succession on human cadavers*.

	Fauna	State of corpse	Approx. age of corpse
A. Exposed corpses			
1st Wave	*Calliphora vicina* (Dipt., Calliphoridae)	'Fresh' (variable with season)	
	C. vomitoria (Dipt., Calliphoridae)		
	Lucilia spp. (Dipt., Calliphoridae)		First 3 months
	Musca domestica (Dipt., Muscidae)		
	M. autumnalis (Dipt., Muscidae)		
	Muscina stabulans (Dipt., Muscidae)		
2nd Wave	*Sarcophaga* spp. (Dipt., Sarcophagidae) [may occur in 1st Wave]	Odour developed	
	Cynomya spp. (Dipt., Calliphoridae)		
3rd Wave	*Dermestes* (Col., Dermestidae)	Fats rancid	
	Aglossa (Lep., Pyralidae)		
4th Wave	*Piophila casei* (Dipt., Piophilidae)	After butyric fermentation protein of 'caseic' fermentation	3–6 months
	Madiza glabra (Dipt., Piophilidae)		
	Fannia (Dipt., Fanniidae)		
	Drosophilidae (Dipt.)		
	Sepsidae (Dipt.)		
	Sphaeroceridae (Dipt.)		
	Eristalis (Dipt., Syrphidae)		
	Teichomyza fusca (Dipt., Ephydridae)		
	Corynetes, Necrobia (Col., Cleridae)		
5th Wave	*Ophyra* (Dipt., Muscidae)	Ammoniacal fermentation Evaporation of sanious fluids	
	Phoridae (Dipt.)		
	Thyreophoridae (Dipt.)		
	Nicrophorus (Col., Silphidae)		4–8 months
	Silpha (Col., Silphidae)		
	Hister (Col., Histeridae)	Remaining body fluids now absorbed	
	Saprinus (Col., Histeridae)		
6th Wave	Acari		6–12 months
7th Wave	*Attagenus pellio* (Col., Dermestidae)	Completely dry	
	Anthrenus museorum (Col., Dermestidae)		
	Dermestes maculatus (Col., Dermestidae)		1–3 years
	Tineola biselliella (Lep., Tineidae)		
	T. pellionella (Lep., Tineidae)		
	Monopis rusticella (Lep., Tineidae)		
8th Wave	*Ptinus brunneus* (Col., Ptinidae)		3 years plus
	Tenebrio obscurus (Col., Tenebrionidae)		
B. Buried corpses			
1st Wave	*Calliphora* & *Muscina stabulans*		
2nd Wave	*Ophyra*		
3rd Wave	Phoridae (*Conicera* may appear on surface)		1 year
4th Wave	*Rhizophagus parallelocollis* (Col., Rhizophagidae)		
	Philonthus (Col., Staphylinidae)		2 years

* Based on Mégnin (1894) as tabulated by Johnston & Villeneuve (1897) and updated by Smith (1973) with further modifications.

1. *Initial decay stage* (0–2 days). Carcase appears fresh externally but is decomposing internally due to the activities of bacteria, protozoa and nematodes present in the animal before death.
2. *Putrefaction stage* (2–12 days). Carcase swollen by gas produced internally, accompanied by odour of decaying flesh.
3. *Black putrefaction stage* (12–20 days). Flesh of creamy consistency with exposed parts black. Body collapses as gases escape. Odour of decay very strong.
4. *Butyric fermentation stage* (20–40 days). Carcase drying out. Some flesh remains at first and cheesy odour develops. Ventral surface of body mouldy from fermentation.
5. *Dry decay stage* (40–50 days). Carcase almost dry; slow rate of decay.

Bornemissza established that the original fauna plays little part in decomposition and emphasised the importance of differentiating between scavengers, predators and parasites among the invading fauna. He also studied the long term effects of carrion on the soil fauna.

There is broad general agreement in the observations of Mégnin (1894), Bornemissza (1957), Reed (1958) and the series of publications by Payne and his co-authors (1965–72). Comparison of Table 1 with Figures 2–5 shows that the same broad groups (orders and families) of insects follow in sequence, although of course the species lists vary with the region. Thus, in the USA the blowfly sequence may be *Lucilia, Callitroga, Calliphora* and *Cynomyopsis*, then *Sarcophaga*, each separated by a period of a few days (Rodriguez & Bass, 1983). The interrelationships between scavenger, predator and parasite in different areas and in different countries are shown in Figures 5–7. The rate of decay will of course vary with temperature, season and micro-climate, and other environmental factors (Tables 3, 4). Species lists will differ according to the region and whether the corpse is in light or shade, and vertebrate scavengers may play a larger role in the tropics. Nevertheless, the careful collection of data on site, coupled with biological knowledge of the insects concerned, will enable a reconstruction of events to be made including an estimate of time of death. Relevant biological details are given in the taxonomic sections of this book. See also Foreman & Smith (1917) and Hepburn (1943).

If corpses are buried in soil or immersed in water the rate of decay and the associated fauna will be different and these and other factors are considered below.

Buried corpses

Mégnin (1887, 1894) and Motter (1898) were the first to discover that exhumed cadavers had their own peculiar insect faunas and later work by Schmitz (1928), Chu & Wang (1975) and Leclercq (1975) confirmed their findings. Burial hinders the decay process by excluding airborne bacteria and the normal faunal succession of invertebrates. However, a limited and somewhat different fauna does get to the corpse; this varies with the nature and depth of the burial.

Table 2 Total number and percentage of species attracted in abundance to the various stages of decay (from Payne, 1965).

Stage of decomposition	Total no. of species attracted to each stage of decomposition	Percentage of species attracted to another stage of decomposition				
		Fresh	Bloated	Active decay	Advanced decay	Dry
Fresh	17	100	94	94	76	0
Bloated	48	33	100	100	90	2
Active decay	255	6	19	100	98	13
Advanced decay	426	3	10	59	100	38
Dry	211	0	1	16	76	100

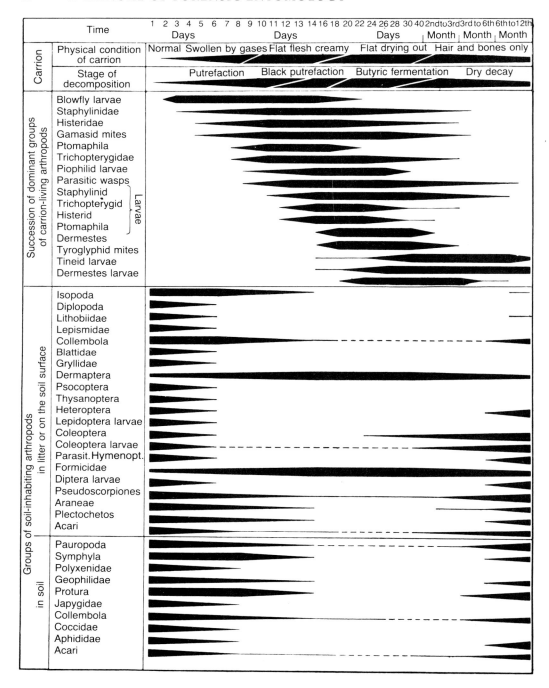

Fig. 3 Arthropod succession on guinea-pig carcases in Australia (Perth) and the effects of carrion decay on the soil fauna. Thickness of bands indicates relative abundance of each group at different times (after Bornemissza, 1957).

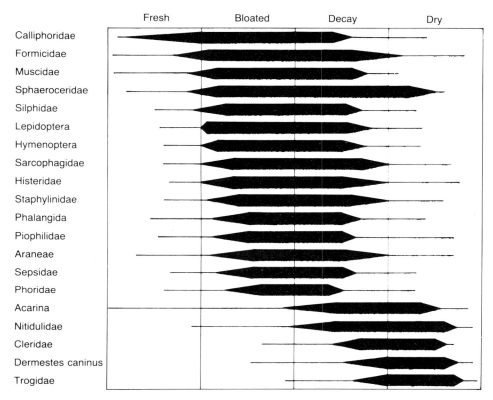

Fig. 4 Arthropod succession on dog carcases in the USA (Tennessee). After Kauppala (*in* Nuorteva, 1977) from data presented by Reed (1958).

A covering of soil will inhibit some but not all of the carrion fauna attracted to exposed corpses (Table 1, Fig. 8; Case history 9, p. 59). Thus, the most significant feature of buried carrion is the longer time taken for its reduction (Figs 9, 10). Payne *et al.* (1968a) found that the reduction of buried pig carcases to 20% of their original weight took six to eight weeks, whereas similar exposed carcases were reduced to 10% of their original weight in one week (Fig. 10). The work of Fourman (1936, 1938) and Lundt (1964) in Germany confirms this.

Table 3 Seasonal variations in the duration (in days) of the decomposition stages of dog carcases in two localities in Tennessee (from Reed, 1958).

Season	Locality	Fresh	Bloated	Decay	Dry
Spring	Woods	4	7	26	90
	Pasture	2.5	4.5	13	75
Summer	Woods	0.9	3.5	11	50
	Pasture	0.9	2.5	10	50
Autumn	Woods	7	9	34	80
	Pasture	2.5	11.5	21	80
Winter	Woods	25	20	95	?
	Pasture	8	32	85	?

Table 4 Mean duration of development (in days) [egg to adult] of some blowfly species cultured on fish under field conditions (from Nuorteva, 1977)*.

	Forest on mainland shore	Island with some birches	Treeless islet with large bird colony	Treeless rock island by open sea	Means for whole area
Lucilia sericata	–	38.2 (28)	26.9 (23)	33.5 (28)	32.8 (23)
L. silvarum	30.5 (28)	25.6 (23)	25.0 (23)	–	28.9 (23)
L. illustris	32.1 (23)	27.2 (23)	31.3 (23)	27.2 (23)	32.1 (23)
L. caesar	36.0 (28)	–	–	–	36.0 (28)
Cynomya mortuorum	31.5 (28)	27.9 (23)	26.2 (23)	25.4 (18)	26.2 (18)
Calliphora uralensis	–	39.0 (39)	–	37.1 (28)	37.2 (28)
C. vicina	33.6 (28)	34.7 (28)	–	31.4 (28)	33.8 (28)
C. vomitoria	38.0 (35)	–	–	–	38.0 (35)

*Experiments extending from 1 July to 15 August 1967 were conducted in the south-western archipelago of Finland (parishes Bromary and Hitis). Mean temperature of nearest meteorological station (Tvärminne) is 15.0°C. The minimal duration of development is given in parentheses.

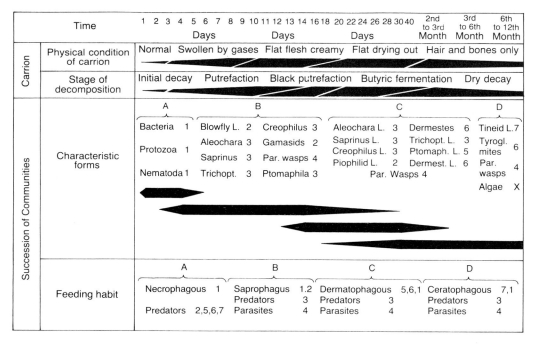

Figs 5, 6 Interrelationships of the carrion fauna. 5, on guinea-pig carcases in Australia (Perth); numbers beside characteristic forms refer to the feeding habit (see bottom part of diagram) and variations in the width of each band indicate the relative activity of each community at different times (from Bornemissza, 1957). 6, on dog carcases in USA (Tennessee); various 'trophic levels' are represented by carrion (rectangle 1), necrophagous species (rectangle 2), omnivores (rectangle 3) and predators (rectangle 4) (? = feeding habit uncertain) (after Reed, 1958).

Some insects lay eggs on the soil surface and the newly hatched larvae reach the carrion by burrowing down through the soil. Such species include the Diptera *Muscina* (Muscidae) and *Morpholeria kerteszi* (Heleomyzidae). The adults of other insects burrow down through the soil and oviposit on the corpse, e.g. Rhizophagidae, Staphylinidae (Coleoptera) and Phoridae (Diptera). The adults of parasitic Braconidae and Proctotrupidae (Hymenoptera) have also been found at depths of at least 50 cm (Lundt, 1964). Blowflies, the main element of the fauna of exposed corpses, may be completely excluded by burial under only 2.5 cm thickness of soil. Other members of the carrion fauna may be less numerous and appear later.

Lundt (1964) found that the top layers of soil at 2.5 and 10 cm were dominated by larvae of the fly *Muscina*. At greater depths he found the phorid flies *Conicera* and *Metopina* and staphylinid beetles of the genus *Atheta*. The coffin-fly (*Conicera*) may appear on the soil surface above a corpse some 12 months after death and serve as an indicator that a corpse lies beneath (see section on Phoridae).

Nuorteva (1977) records staphylinid beetles of the genus *Philonthus* on buried corpses in Finland. In the USA, Payne *et al.* (1968a) found a different fauna on buried pigs, dominated by ants (*Prenolepis imparis* (Say), phorid flies (*Dohrniphora, Metopina*) and staphylinid beetles (*Aleochara, Oxytelus*). Payne *et al.* (1968a) recognized five waves in the faunal succession on buried pigs, summarized below.

1. *Fresh*. This was the period between death and initial bloating and lasted about three days. Ants, especially *Prenolepis imparis*, fed actively on blood and exposed moist skin at the mouth, abdomen and ears.

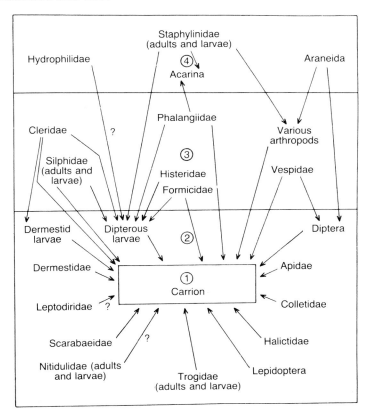

Fig. 6

22   A MANUAL OF FORENSIC ENTOMOLOGY

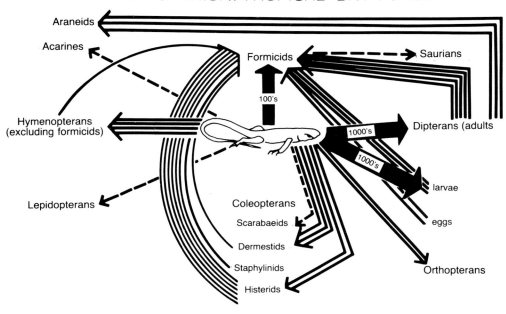

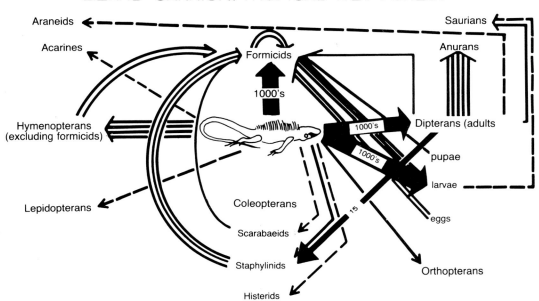

Fig. 7 Food webs of organisms associated with lizard carrion in contrasting tropical habitats. Each normal solid line represents one feeding observation. The heavy solid lines show relative numbers of feeding interactions. Where a species was repeatedly seen on the carcase under observation but not seen to feed, the observation of it having fed at another carcase is shown as a dotted line (after Cornaby, 1974).

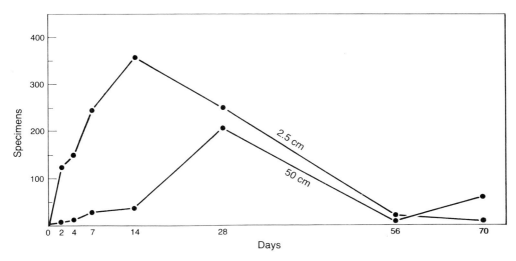

Fig. 8 The development of the insect fauna on buried carcases under soil covers of 2.5 cm and 50 cm. The development of the fauna is much delayed under a thick soil cover, but even a thin soil cover caused considerable delay (after Lundt, 1964).

2. *Bloating*. Ants and the following Diptera (which arrived on the third day) were dominant: *Leptocera* spp. (Sphaeroceridae); *Dohrniphora incisuralis* (Loew) and *Metopina subarcuata* Borgmeier. Psychodidae were observed on the fifth day and by the seventh day, as deflation began, sphaerocerid and phorid larvae were actively feeding.

3. *Deflation and decomposition*. The feeding of ants and maggots apparently determined the time and rate of deflation. Odour very strong. By the tenth day flies and maggots numerous. The beetles *Oxytelus insignitus* Gravenhorst and *Aleochara* spp. feed on small maggots at this stage and cynipid and diapriid parasites (Hymenoptera) reach the corpses. Colonies of fungi and bacteria become established on congealed fluids.

4. *Disintegration*. Maggots and flies numerous and the larvae of Psychodidae, Phoridae and Sphaeroceridae were especially active around the remaining soft tissues. Mites (Acari), springtails (Collembola), Cryptophagidae (Coleoptera), Sciaridae (Diptera) and millipedes (*Cambala annulata*) appeared. The colonies of fungi and bacteria now covered the carcase. The maggots migrated from the carcase at the end of this stage. Between days 30 and 60 mites of the genus *Caloglyphus*, and the Collembola *Folsomia fimetaria* L. and *Hypogastrura armata* Nicolet were the main scavengers.

5. *Skeletonization*. Ants, flies, Collembola and mites now the dominant fauna. Spiders, centipedes and millipedes were also present.

In general terms the processes of decay and the major groups involved in the succession are similar to corpses on the surface. However, the presence of purely surface-type fauna (e.g. blowflies) should lead one to suspect that burial may not have taken place immediately after death.

The entomologist is sometimes asked to examine insects from buried bodies excavated on archaeological sites or from ancient Egyptian tombs (see also Mummification, p. 27). While this is unlikely to have any medico-legal application today it can help to form a picture of burial practices such as cremation before burial, or of how long bodies were left exposed to insects before burial, and some data of forensic interest can be gleaned. Teskey & Turnbull (1979) found the following Diptera puparia in pre-historic (2000–2500 years old) graves in the Augustine Mound, New Brunswick, Canada: 563 *Phormia regina*

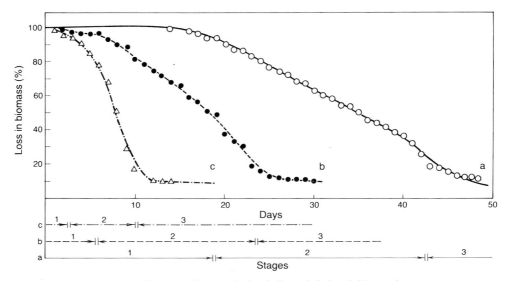

Decomposition rate (points indicate daily loss in biomass)
a – spring, underground, b – summer, underground, c – summer, surface; stages: 1 – preparatory, 2 – active decomposition, 3 – residual

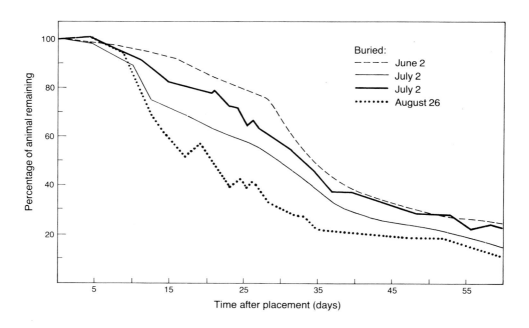

Figs 9, 10 Effects of burial on the decomposition rate of carrion. 9, for bank vole (*Clethrionomys glareolus* Schreber) in Poland (from Nabaglo, 1973). 10, for pig carrion during autumn 1966 in South Carolina, USA (after Payne *et al.* 1968a).

(Meigen), 29 *Protophormia terraenovae* (Robineau-Desvoidy), 1 *Cynomyopsis cadaverina* (Robineau-Desvoidy), 14 *Muscina assimilis* (Fallén), 2 *Hydrotaea* and 12 Heleomyzidae. Clearly the bodies had been exposed for a considerable time before interment, though sufficient flesh remained to attract some species after burial (e.g. *Muscina* and Heleomyzidae).

Corpses immersed in water

The following details of human decomposition are from Simpson (1985). Decomposition is retarded in water because body heat is lost twice as fast as in air.

In water the head sinks low and blood gravitates towards the head and neck, so that decomposition begins there. Thus the greenish discoloration of the flanks may not appear until five or six days after death and in average weather conditions the generation of gases may not cause the body to float until 6–10 days after death. The skin of the hands and feet becomes wrinkled after 10–12 hours and after about ten days may readily peel off if serious decomposition has not occurred. Also, by this time the hair becomes loose and 3–4 weeks after death the finger and toe nails are easily detachable. These periods may be halved in summer conditions. Finally, the flesh breaks down into slime and the skeleton collapses.

Simpson also gives valuable information on the fate of ectoparasites of potential forensic value. Fleas are drowned in about 24 hours but if immersed for only 12 hours they require about an hour to revive and after 18–20 hours immersion, a period of some four to five hours. Body lice usually die in 12 hours following immersion. Blowfly larvae already present on a body when immersed will not survive for long and if still alive may thus indicate recent removal from another site (Aruzhonov, 1963). Pollution of water by sewage or chemicals or warm water near power stations etc. could modify the fauna and affect rates of development, etc.

The fauna on totally immersed corpses will obviously differ from that on exposed or buried corpses, but if part of the body projects from the water (or was exposed for a while before immersion) blowflies and other 'surface' insects may be present (see Case history 17, p. 65). The fauna will also vary between fresh and salt water. Mégnin (1894) reports cirripede Crustacea on an immersed corpse, the size of which would indicate the time of immersion of the corpse. Other Crustacea, such as shrimps, prawns and probably crabs, will feed on submerged corpses in marine situations. Holzer (1939) records the larvae of caddis-flies (Trichoptera) feeding on a foetus immersed in water for 24 hours; the damage done was already considerable (see Fig. 380, p. 163).

The only detailed study of immersed carrion is by Payne & King (1972) working with pig carcases in the USA (Figs 11, 12). They recognized six stages as follows:

1. *Submerged fresh*. A few pigs floated when first immersed, but the majority sank. The pigs bloated and began to float after 1–2 days in summer, but remained submerged for 2–3 weeks in the autumn and winter. Only hydrophilid beetles occurred on carcases at this stage.

2. *Early floating*. The distended abdomen projected above the water first and was immediately covered in blowfly eggs. The principal blowflies laying on all exposed parts of the carcases were *Lucilia caeruleiviridis* (Macquart) and *Cochliomyia macellaria* (F.). Payne & King found that many more flies oviposited on these pigs than on carcases exposed on the ground. They also recorded that many other flies and wasps congregated, but only blowflies and fruit-flies (Tephritidae) stopped to feed on the pigs. Wasps and hornets preyed upon the blowflies (adults and eggs) feeding on the pig juices. Decay odour was pronounced and gas bubbles were evident on the water surface.

3. *Floating decay*. Blowfly eggs had hatched by the third day (c. 23 hours after laying on the bloated carcases). Many openings were created in the skin by the intense feeding of

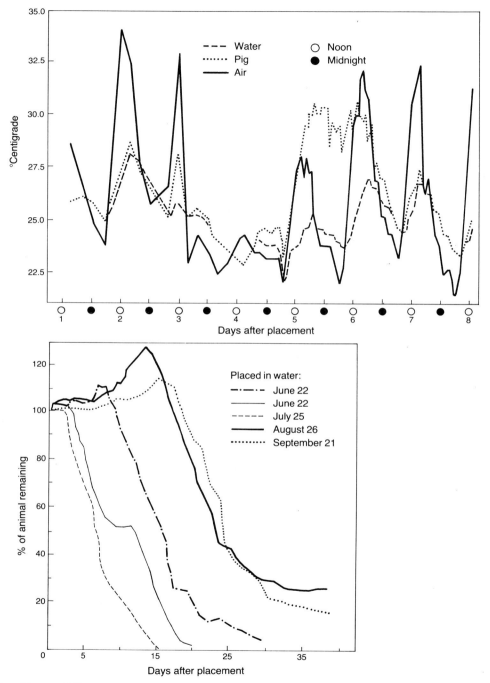

Figs 11, 12 Effects on pig carrion when submerged in water in South Carolina, USA. 11, comparison of air, water and pig temperatures; 12, decay curve showing loss of weight of submerged pig carrion in the summer and autumn of 1966 (after Payne & King, 1972).

small larvae which could not gain entrance to the normal orifices which were submerged. Blowfly oviposition had ceased by the third day and the beetles *Necrodes surinamensis* (F.) (Silphidae) and *Staphylinus maxillosus* L. (Staphylinidae) fed on the blowfly maggots. The odour of decay resembled that of fetid cucumbers and at night staphylinid and histerid beetles were abundant on the corpses.

4. *Bloated deterioration.* Maggot activity intense, many maggots and predatory beetles were forced into the water as the exposed surface of the pig reduced. By the seventh day most exposed tissues were gone and the maggots had migrated. Histerid, staphylinid and silphid beetles were preying on the mass of maggots.

5. *Floating remains.* Few maggots and histerid beetles, sphaerocerid, phorid, drosophilid and psychodid flies feeding on the remains. Many dead maggots on the water. This stage lasted 4–14 days and ended when remains sank.

6. *Sunken remains.* This stage lasted 10–30 days when only bones and bits of skin remained. Decomposition completed by bacteria and fungi. Mosquito larvae, *Culex pipiens quinquefasciatus* (Say), were abundant in the foul water. Adult flies – Psychodidae, Sphaeroceridae, Chloropidae – were presumably feeding on the dead maggots and floating carrion fragments.

In all, these authors collected 102 insect species from 37 families, of which 93 had been found in studies on dead pigs on the soil surface.

In natural conditions and not in an artificial tank the fauna would probably have been much richer and have included many more truly aquatic insects with carnivorous larvae. Such a study would be well worth while.

Mummification

If the temperature is sufficiently high, and/or there are suitable air currents, bodies may become mummified and putrefaction be prevented by the exclusion of bacteria and most or all of the normal carrion fauna. Bodies hidden indoors in chimneys, cupboards or under floorboards may become mummified. New born babies, which are sterile and less prone to decomposition, are the commonest human mummies likely to come to forensic notice. A state akin to mummification can also occur in low temperatures (Fig. 13).

When the body is dry it may be invaded by the stored product type of insects, frequently occurring indoors, including beetles (such as *Dermestes*), clothes-moths or mites.

Ancient Egyptian mummies have also been examined for insects and may include beetles of the families Tenebrionidae, Ptinidae, Anobiidae and Dermestidae. Diptera recorded include *Chrysomya albiceps* (Wiedemann), *Piophila casei* (L.) and *Musca domestica* (L.).

Bodies that have been kept indoors in dry conditions and not available to the early waves of blowflies, but are not actually mummified, may become infested by maggots of *Ophyra* (Muscidae) after a period of several months (see Case history 9, p. 59).

Burnt bodies

There is little detailed information on the effect of burning on the faunal succession in bodies. Wardle (1921) notes that not only must the surface of the flesh be moist to attract ovipositing blowflies, but the protein content should not be coagulated by heat. He found that slices of raw liver, if superficially singed with a Bunsen flame or allowed to desiccate naturally in the sun, were no longer attractive for oviposition even if re-moistened. He also found that freshly cooked meat, however moist, is not attractive.

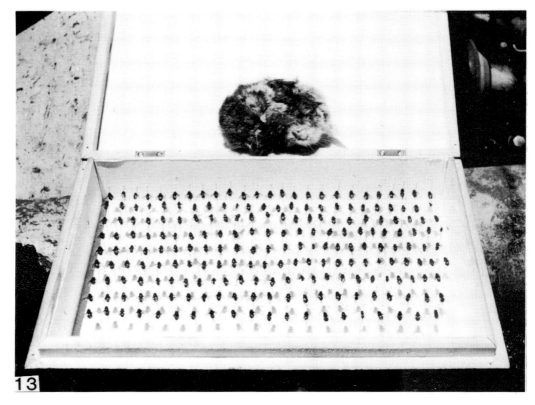

Figs 13, 14 Blowflies and carrion in Arctic conditions. 13, blowflies, *Boreellus atriceps* (Calliphoridae), all reared from one dead lemming; due to the low temperatures there was little putrefaction and the skin of the animal remained intact (Photo: J. A. McAlpine). 14, the nutrients in the soil under the remains of a dead musk ox cause a more luxuriant growth of vegetation (Photo: David Bellamy).

Nuorteva *et al.* (1967) found *Calliphora vicina* Robineau-Devoidy and *Fannia canicularis* L. larvae in the ears, eyes and mouth of a badly burnt body (see Case history 14, p. 63) and I have seen *Calliphora vomitoria* L. and *Lucilia caesar* L. from burnt bodies in Britain, but on a burnt foetus only *Ophyra* and *Fannia* were present. Hopkins (1944) records *Lucilia cuprina* (Wiedemann) from an arm burnt by lightning and an ear injured by burns in Uganda.

Geographical location

Clearly the zoogeographical region (Fig. 46), country and type of terrain within a given country will affect the composition of the faunal succession and the rate of decomposition (Tables 3, 4). In polar conditions the available carrion fauna may be drastically reduced in numbers of species. Thus on dead lemmings on Ellef Ringnes Island in Arctic Canada, McAlpine (1965) found only one species of blowfly, *Boreellus atriceps* Zetterstedt. This circumpolar species was the only blowfly seen on the island, where it was also found on corpses of the Arctic hare and a husky dog. No doubt it would have found a human corpse equally suitable for development. Due to climatic conditions there was little or no

putrefaction and large numbers of the blowfly were reared from an individual lemming (Fig. 13). The skins of the animals remained intact after emergence of the flies and the 'mummified' bodies retained their shape though all the flesh had gone. Another interesting environmental response was that unlike most Calliphoridae the larvae of *B. atriceps* did not burrow under or leave the carcase to pupate. Instead they moved to the upper surfaces of the corpse, seeking the warmth of the direct rays of the sun.

A fascinating possibility of forensic applications in arctic or subarctic regions struck me forcibly when watching a lecture by Dr David Bellamy, in which he showed a slide (Fig. 14) of what appeared to be a grave of lush vegetation amid the barren wastes on the site where a musk ox corpse had been. This was a quite natural phenomenon. The simple explanation, of course, was that the nutriments had drained from the corpse into the impoverished soil and had favoured the development of flowering plants. Such a sight would provide an obvious clue to the whereabouts of a large corpse in these regions. The papers by Nuorteva (especially 1965, 1966a, 1966b, 1972) and Hanski & Nuorteva (1975) describe work carried out in arctic or subarctic conditions.

In temperate regions the range of the carrion fauna increases and in Europe about 38 species have been associated with human corpses, 56 with foxes and 36–38 with rabbits. In South Carolina (USA) Payne found 306 species on pig carcases. In Japan, Suenaga (1959) recorded numbers of flies reared from rat, fish, snake, chicken and frog (descending in that order) but he only used heads.

In more tropical regions the carrion fauna may not be richer in species, but groups not normally important on carrion (e.g. ants, cockroaches) or peculiar to tropical regions (e.g. termites) may play a role in its reduction. Seasonal variation in the fauna is less evident in the tropics and blowflies may breed throughout the year (Cornaby, 1974; Jirón, 1979; Jirón & Cartín, 1981).

Cornaby (1974) made a comparative study of toad and lizard corpses in tropical dry forest and tropical wet forest in Costa Rica. He found 170 species associated on carrion, of which the Calliphoridae, Sarcophagidae (Diptera), Formicidae (Hymenoptera) and Scarabaeidae (Coleoptera) were most important in reducing carcases to the dry decay stage. He studied the interactions of necrophagous animals and their predators, temporal succession patterns and species compositions between sites (Fig. 7).

However, on elephant carcases in Kenya, M. Coe (1978) found only a small fauna, comprising two blowflies (*Chrysomya*), a few ants, beetles and a termite. Although there were only two species of blowflies on the carcases the numbers of individuals were high. Coe estimated that there were 14 460 larvae of *C. albiceps* Wiedemann and 11 500 *C. marginalis* Wiedemann (= *regalis* R.-D.) around a single elephant carcase. The *C. albiceps* larvae became important predators of the *C. marginalis* larvae in the migratory phase.

Temperature and humidity

The major factors controlling oviposition and the rates of development of sarcosaprophagous insects are temperature and humidity. Cold weather and rain inhibit fly activity. Nagasawa & Kishino (1965) have established that Pradhan's (1946) formula correlating the rate of insect development with temperature is substantially applicable to Diptera (Fig. 15). Reiter (1984), however, also found that while maximal larval length was reached earlier at higher temperatures, once reached it again declined and the decrease in length became more rapid at higher temperatures (Fig. 16). He also found that constant temperatures over 30°C led to stunted forms which failed to pupate and died, and that constant temperatures under 16°C induced a state of rest once maximum growth had been attained and metamorphosis resumed again when the temperature was raised. Thermal death points for insects range from $-15°C$ to $-30°C$ (Knipling & Sullivan. 1957) with an upper limit at about 60°C (Knipling, 1958), but there is wide variation between species.

Forensic scientists have in the past frequently dismissed entomological forensic methods as too temperature dependent. However, this comment applies equally to all biological methods of estimating time of death. Clearly temperature and humidity will vary with the season, geographical and topographical location, and between micro-habitats within the site in which a corpse is found (Nuorteva, 1972; Hanski, 1976a, 1976b). It is the daily range of temperature at the site in which a corpse is found that concerns the forensic investigator and some indications of this and its effect on rate of development for different seasons are given in Figures 17 and 18. Deonier (1940) found that the minimum temperatures for activity of a variety of blowflies about carcases ranged between 40 and 60°F.

Nielsen & Nielsen (1946) found that *Calliphora* eggs failed to hatch below 4°C but at 6–7°C both hatching of eggs and larval development occurred. Ideally, the temperatures of the corpse and its surroundings should be taken on the site, but this rarely happens in practice and the forensic worker usually has to interpolate these data retrospectively. However, if records of air temperatures can be obtained from the nearest meteorological station it may be possible to interpolate and estimate relevant temperatures for the particular micro-climate under investigation (see also sections on *Calliphora vicina* Robineau-Desvoidy and Procedure in the Laboratory). Such interpolation must be made with care and may be misleading (Aruzhonov, 1963; Leclercq & Tinant-Dubois, 1973; Leclercq & Watrin, 1973; Nuorteva, 1972). A comparison of carrion temperature with soil temperature during each stage of decomposition is shown in Figure 19. Deonier (1940) found that carcases were at a higher temperature than the air, partly due to the effect of the sun but principally from heat produced by the maggots.

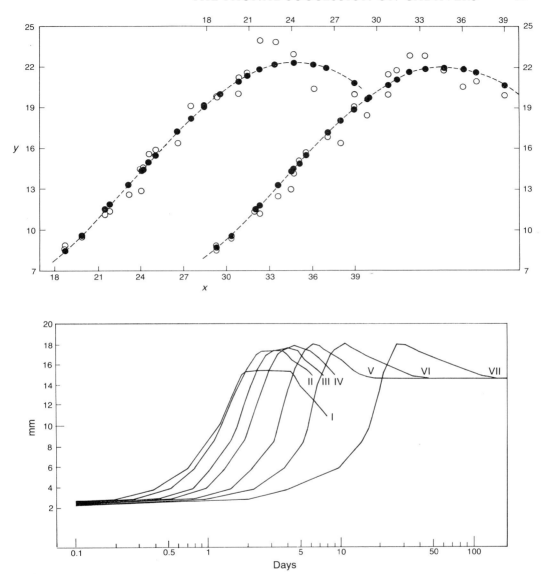

Figs 15, 16 Effects of temperature on rate of development. 15, for puparia of *Musca domestica*; upper abscissa for males, lower abscissa for females, open circles denote the observed, solid circles denote the calculated values respectively (from Nagasawa & Kishino, 1965). 16, development rates of larvae of *Calliphora vicina* at constant temperatures, I = 35°C, II = 30°C, III = 22–23°C, IV = 18–19°C, V = 14–16°C, VI = 10–12°C, VII = 6.5°C (after Reiter, 1984).

Light and shade

Some insects prefer light, i.e. they are positively phototropic (or if the sun is the light source, positively heliotropic), or may shun the light, i.e. they are negatively phototropic

32   A MANUAL OF FORENSIC ENTOMOLOGY

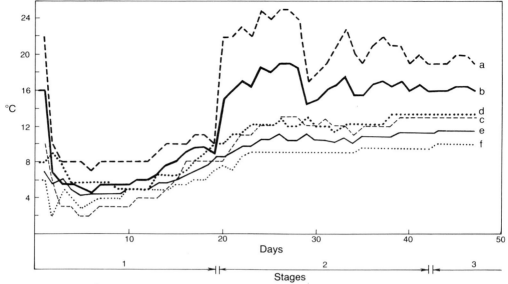

Daily fluctuations in temperature during spring underground decomposition
Temperatures on surface of ground: a – maximum, b – average, c – minimum
Temperatures underground: d – maximum, e – average, f – minimum; 1, 2, 3, see Fig. 9

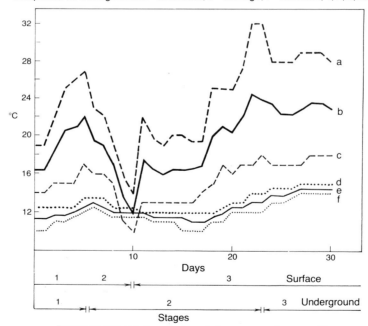

Daily fluctuations in temperature during summer decomposition
Temperatures underground: d – maximum, e – average, f – minimum; 1, 2, 3, see Fig. 9
Temperatures on surface of ground: a – maximum, b – average, c – minimum

Figs 17–18 Daily fluctuations in temperature during decomposition of bank vole carcases in Poland. 17, during spring underground decomposition; 18, during summer decomposition (after Nabaglo, 1973; see also Fig. 9).

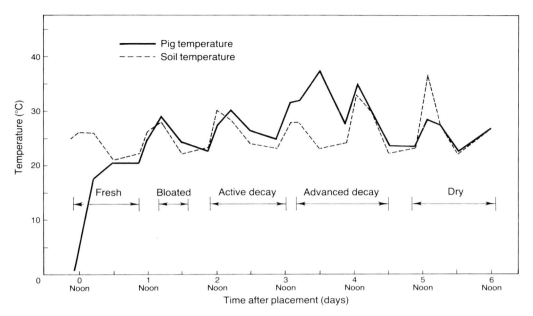

Fig. 19 Comparison of average exposed pig carrion temperature with soil temperature during each stage of decomposition (after Payne, 1965).

(or heliotropic). This may affect the particular insects present on a corpse (and will also affect temperature). It is well known that of the two commonest genera of blowflies associated with carrion, bluebottles (*Calliphora*) prefer shady conditions and greenbottles (*Lucilia*) prefer sunlight. The fleshflies (*Sarcophaga*) also prefer sunlight. Smith (1956) found that moving specimens of the carrion-smelling fungus *Phallus impudicus* alternately into conditions of light or shade profoundly affected the numbers of these three genera visiting the fungus.

Thus, on bodies found indoors one would normally expect to find *Calliphora*, but not *Lucilia* or *Sarcophaga* (unless all the windows were open and the body in bright sunlight). Similarly, out of doors the maggots present may indicate whether the body has been lying in bright sunlight, deep shade, or perhaps both, at different times of day (see also the taxonomic sections under the individual genera).

Seasonal and daily periodicity

Different insects occur at different times of year or day and are said to have different flight periods. A corpse exposed in the spring and summer will have a richer and different fauna from one exposed in the winter when the faunal succession is absent. The commonest blowflies have a number of generations per year and therefore a fairly long flight period, from early spring to late autumn, though they may not be reproductively active throughout the whole of that period (except in the tropics). Corpses exposed during the winter, when no blowflies are about, will have a smaller fauna consisting largely of ground level insects such as beetles. Little work has been done on carrion fauna in winter conditions and it was only recently discovered (Broadhead, 1980; Erzinçlioğlu, 1980) that the larvae of winter gnats (Diptera, Trichoceridae) then feed on carrion. There is scope for research here. Carrion insects also have daily (diel) rhythms of activity (Lewis & Taylor, 1965; Norris, 1965*b*; Suenaga, 1963; Hanski & Nuorteva, 1975).

Availability of food and competition

It should be borne in mind that this book is concerned with the *invertebrate* fauna found on a corpse in the course of a forensic investigation. Clearly, vertebrate scavengers may be the first to arrive at a corpse and indeed may even have been the cause of death, especially in more exotic tropical environments. However, vertebrates have little forensic importance in association with cadavers unless they move the corpse or at least part of it. They have occasionally led to the discovery of corpses, e.g. dogs locating bodies by smell and foxes seen carrying bits of a body in polythene bags! Hewson & Kolb (1976) discuss scavenging of carcases (of sheep) by foxes and badgers (see also Akopyan, 1953). The breaking up of carrion by vertebrates enhances desiccation which may affect the invertebrate fauna (Hancox, 1979).

The size of the corpse may limit the availability of food and in turn affect the survival of the insect larvae feeding upon it (Suenaga, 1959; Denno & Cothran, 1975; Kuusela & Hanski, 1982). However, shortage of food need not prevent completion of the life history though it may result in small larvae, pupae and adults, and thus give a misleading estimate of age. This may happen among the later stages or generations of early invaders, even on initially large corpses when most of the flesh has been removed or dries out very quickly.

The important factor affecting size or even completion of the life history of carrion insects is *competition* for the food source. This occurs in three ways:

1. Intraspecific competition (i.e. within a species) reduces the size of the larva and subsequently the number and fecundity of individual adults

2. Interspecific competition (i.e. between species) with similar results to (a) plus a possibility of total elimination

3. Predators and parasites – selective predation or parasitization of one species will be advantageous to a competing species

Early arrivals ovipositing on a corpse (such as *Calliphora* and *Lucilia*) will clearly have an advantage, but some later arrivals such as *Chrysomya* and *Sarcophaga* seem to compensate for this by being viviparous and deposit living larvae on the carcase. However, if food is scarce even *Calliphora* females may retain eggs in their bodies until they hatch and then deposit the living larvae.

Some Diptera larvae may be carrion feeders at first and become predators in the second and third instars (e.g. *Hydrotaea* and *Ophyra*). M. Coe (1978) found that, of the larvae of two species of *Chrysomya* feeding on elephant carcases in Kenya, the larger 'spiky' larvae of *C. albiceps* became predaceous on the smaller unarmed *C. marginalis* only in the migratory phase, as they left the carcase (see also under *Chrysomya*). Competition between predatory and carrion-feeding beetles and blowfly larvae may be intense (see Coleoptera section, p. 138). Parasitism of blowfly larvae by the braconid *Alysia manducator* (Hymenoptera) can stimulate early pupation (see also Hymenoptera section, p. 154).

A detailed discussion of competition is given by Varley *et al.* (1973) and for its particular significance on carrion Fuller (1934), Denno & Cothran (1975) and Putman (1983) should also be consulted (see also sections on Coleoptera (p. 138) and Acari (p. 165).

Manner of death

The manner of death can affect the rate of decomposition and the faunal succession. Death by gassing, especially from carbon monoxide (e.g. coal gas, car exhaust fumes), causes changes in the blood haemoglobin which affect the body tissues. Utsumi *et al.*

(1958) studied rats killed in nine different ways – aeroembolism, arteriotomy, drowning, burning, hydrocyanic potassium poisoning, arsenic acid poisoning, parathion poisoning, phosphorus poisoning and poisoning with hypnotics (bromdiethylacetyl urea). Poisons such as parathion and arsenic prolonged the rate of decomposition and affected the rate of growth of beetles and flies feeding on the carrion, so that the cadaver fauna, though rich at first, later declined. Beyer *et al.* (1980) were able to identify the drug (phenobarbitol) used in a case of fatal overdose by analysing the maggots from a badly decomposed body on which insufficient tissue was available for study.

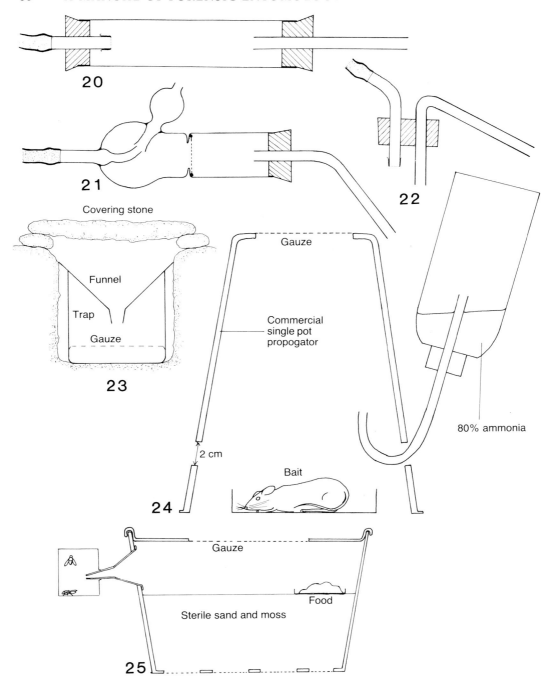

Figs 20–25 Apparatus for collecting, trapping and rearing carrion insects. 20–22, aspirators or 'pooters'; 23, ground trap; 24, Lane-type baited carrion trap; 25, Nuorteva type rearing container.

Methods and techniques

Equipment for collection and preservation

The following basic equipment is essential:

1. Rubber gloves and occasionally other protective clothing.

2. Forceps for picking specimens off a corpse individually or a small potting trowel could be used for larger samples.

3. A small artist's paint brush is useful for picking up tiny specimens that might be damaged by forceps.

4. Insect aspirators or 'pooters' (Figs 20–22) normally used for collecting small insects are not really suitable for collecting off corpses due to the health hazard. However, pooters are useful for collecting small insects and larvae from soil samples, but if one is used it should be of the blow variety (Fig. 21).

5. Nets of the kite or folding variety (Fig. 26–28) are useful for collecting insects flying over or settled on a corpse and a pond net (Fig. 29) should be used for collecting off corpses in water.

Specimens collected should be placed alive in glass or plastic tubes or jars, preferably with corks or cap that can be easily removed as the tube is held in one hand, leaving the other hand free for holding a net or forceps.

A medium sized jar with a good stopper or cap can be used for killing adult specimens by dropping them in with a small ball of cotton wool soaked in a suitable killing agent, such as ethyl acetate, 0.880 ammonia or carbon tetrachloride. The usual safety precautions should, of course, be taken regarding danger from inflammability or inhalation. Specimens can then be transferred to 70–80% ethyl alcohol. Formalin is not recommended.

Few forensic workers will want to pin specimens unless they are anxious to form a reference collection, when the literature cited at the end of this section should be consulted. If male adult flies are pinned for sending to a specialist it is essential to draw out the genitalia with a needle so that the details can be seen with ease after the fly is dry. Genitalia are widely used as taxonomic characters by specialists, especially in *Lucilia* and *Sarcophaga*, but are ignored in the keys presented in this book, since it is intended for non-specialists. If a large number of adult insects are collected for posting to a specialist they can be carefully layered between strips of cellulose wool cut to size and packed in a small, but strong box (Fig. 30).

Insect larvae can be killed by dropping them into near-boiling water and then transferred into 70–80% alcohol or Pampel's fluid. The latter is particularly suitable for Diptera larvae requiring dissection. Whenever possible specimens should be placed in Pampel's fluid alive as this improves the preservation of internal structures. There are various formulae for Pampel's fluid but the following is recommended:

Formalin (35%)   6 parts
Ethyl alcohol (95%) 15 parts
Glacial acetic acid 2 parts
Distilled water   30 parts

The fluid should be mixed and used cold. For killing and fixing larvae, etc. immerse in the fluid while alive and leave to harden for up to 2–3 weeks, then remove and store in 80% alcohol.

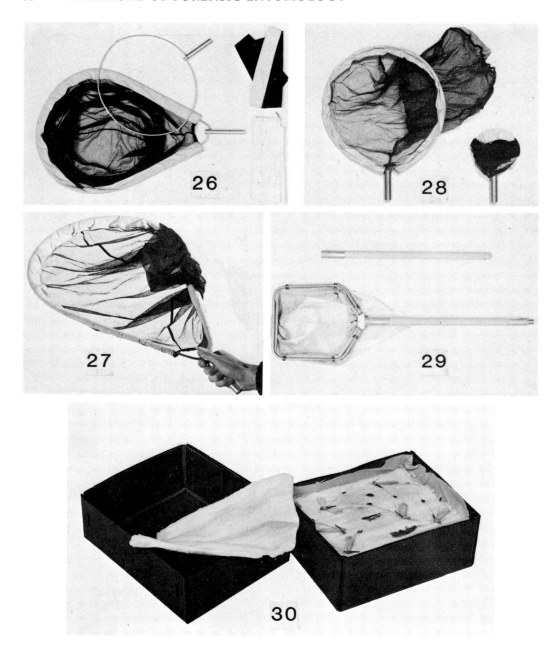

Figs 26–30 Collecting apparatus. 26–27, aerial kite net; 28, aerial folding pocket net; 29, pond net; 30, insects layered between strips of cellulose wadding in postal box.

For subsequent microscopic examination the following techniques used by colleague Dr K. M. Harris in his work on the very delicate gall midges (Cecidomyiidae) are particularly recommended. Preserved Diptera larvae may need to be 'cleared' so that the detailed structure of the mouthparts and spiracles (anterior and posterior) can be studied. This can be achieved by macerating the specimen in a 10% aqueous solution of KOH (caustic potash) for at least 15 minutes in a watchglass or small dish. Specimens that have been in alcohol for six months or more may need a longer period of maceration, e.g. up to 12 hours in cold KOH or a shorter period in warm KOH. Specimens should then be transferred to a small amount of glacial acetic acid for at least 15 minutes to neutralise any residual KOH. Then an equal amount of Berlese preservative should be added and the container covered with a lid and left for 24 hours, by which time the specimens should be slightly distended and cleared. If the specimens are not clear then they should be soaked in water for 2–3 hours and the whole process repeated. Specimens can then be examined, mounted on slides or stored in the preservative permanently. Berlese preservative should be made up as follows:

| Chloral hydrate | 20g | 50% w/w glucose syrup | 5ml |
| Glacial acetic acid | 5ml | Distilled water | 30–40ml |

This produces a clear fluid which will not set, but if it should thicken due to evaporation small amounts of distilled water can be added as a thinner.

Berlese mountant for slides is made by adding 12 g of gum arabic to the above formula. Permanent slides of specimens can be made in this medium if ringed with Euparal or other suitable medium when the slides are dry (after at least three weeks at 35°C).

Other techniques for the collection and preservation of insect specimens are described by Oldroyd (1970a), Cogan & Smith (1974), Norris & Upton (1974), Martin (1978), Smith (1973a), Stubbs & Chandler (1979) and A. K. Walker & Crosby (1979).

Recent work with the scanning electron microscope has revealed characters of value in the identification of Diptera larvae (Kitching, 1976; O'Flynn & Moorehouse, 1980 and Erzinçlioğlu, 1985). Forensic entomologists having access to such an instrument should study the above papers and the technique suggested by Grodowitz et al. (1982).

Procedure on site

Any insects hovering around or settled on a corpse should be captured with a suitable net, transferred to a glass or plastic tube, killed and preserved in 80% alcohol. Adult insects (or other invertebrates) which are reluctant or unable to fly may be caught by 'persuading' them into tubes with the aid of forceps, dissecting needle, stick or grass stems. If kept alive each specimen should be in a separate tube to avoid possible predation. Eggs, maggots and any other larvae (of all sizes) feeding on the body, especially in the natural orifices, should be collected with forceps or small trowel into 80% alcohol or Berlese's fluid (see previous section) and a small sample of each type and size kept alive for subsequent rearing in the laboratory.

It is important to note the precise site and situation on the body of each sample collected and this should be recorded on the spot. Each type and size of maggot at each site on the body should be taken. Any fly puparia should be kept alive in separate tubes and the colour noted at the time of collection. The puparium is pale yellow immediately after pupation and takes several (c. 24) hours to harden and tan to a dark brown (Fig. 31; Thomsen, 1934; Schumann, 1965). Where puparia are the only stage available it may be important to know if they had only recently been formed when collected, especially if they are examined later by someone else (see Life histories, p. 52, Case history 11, p. 60). Dissection in the laboratory will of course reveal the state of development of the true pupa inside and give another indication of age (Fraenkel & Bhaskaran, 1973). If one wave of maggots has already left the body, puparia and inactive larvae or prepupae (see Life histories, p. 52) should be sought by making a series of shallow transects (3 in

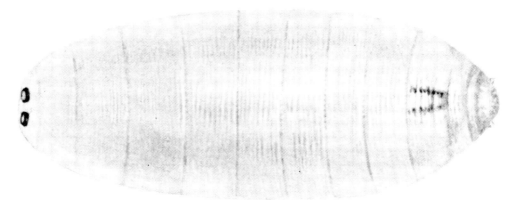
Puparium of *Musca domestica* immediately after pupation

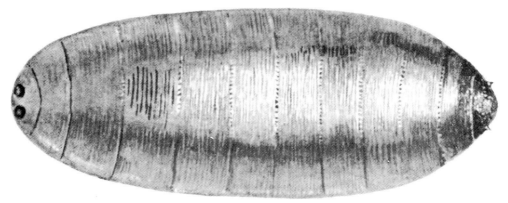
Puparium of *Musca domestica* a few hours old

Older Puparium of *Musca domestica*

Fig. 31 Pigmentation or darkening of the puparium in the higher Diptera takes several hours (from Thomsen, 1934). Since the third larval stage and the prepupal stage (inside the puparium) are the longest developmental stages, the presence of pale freshly formed puparia can be a critical indicator of the time a corpse has been exposed.

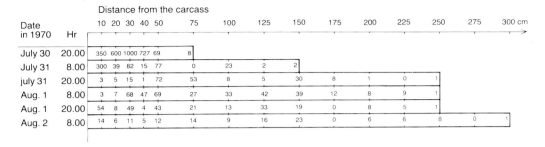

Fig. 32. Dispersal of blowfly larvae from a 'carcase' (7 kg of raw fish) in a lightly wooded area with well-developed moss layer and undergrowth in Finland (Siikainen). Five traps were placed at different distances (see abscissa of diagram) and the numbers of larvae found in the traps are shown. Dispersal commenced five days after exposure of the 'carcase' during a mean temperature of 16.8°C (from Nuorteva, 1977).

(7 cm)) in several different compass directions up to 20 ft (6 m) from the corpse (Fig. 32). Empty puparia indicate that a generation of flies has emerged (see Case nos 14, 15), but since they also contain the third instar mouthparts and spiracles (Figs 39, 40, 43) they can facilitate identification. Great care should be taken to secure all the puparial trappings. On emergence of the fly, the puparium splits anteriorly into two parts under pressure from the fly's ptilinum (an inflatable sac on the head). The final instar larval mouth-hooks are attached to the ventral part of the 'cap' and should be carefully preserved as they can be soaked off (in KOH) in the laboratory for further study.

Puparia of some species may also be found on the body, in the clothing, or beneath the body. The soil beneath the body (and above in buried corpses) should therefore be sampled for any fly larvae or puparia, beetles or other invertebrates

If vegetation has been used to obscure the body, non-carrion insects may have fallen on to the body. Samples of the plants and any insects thereon should be collected separately as they can sometimes yield clues of forensic value (see Case history 11, p. 60).

Samples of any insects present on corpses immersed in water should be collected with a pond net and either preserved in alcohol or brought back alive in watertight containers for subsequent recovery. Any maggots or other insects found on parts projecting above the surface should be collected and suitably labelled.

A full description of the site and condition of the corpse should be made and photographs should be taken. Where possible, records of the physical conditions of temperature, humidity, and soil samples should be gathered. If temperatures cannot be recorded on site then one should ascertain the location of the nearest meteorological station so that records of maximum and minimum temperatures can be obtained later.

Procedure in the laboratory

If the preserved larvae collected on site prove to be Diptera of the third instar (see Life histories, p. 52, and Figs 36–40) or puparia (Figs 31, 41, 42), thus facilitating identification, it may not be deemed necessary to rear the live samples and these too can be preserved in spirit (remember that they have 'aged' since collection and should therefore be treated as a separate sample). However, if larvae are not final instar or are otherwise not readily identifiable (and experience soon dictates which are 'difficult'), they should be reared either on small bits of tissue from the corpse or on a suitable substitute. If the actual field conditions (pabulum, temperatures, etc.) can be reproduced

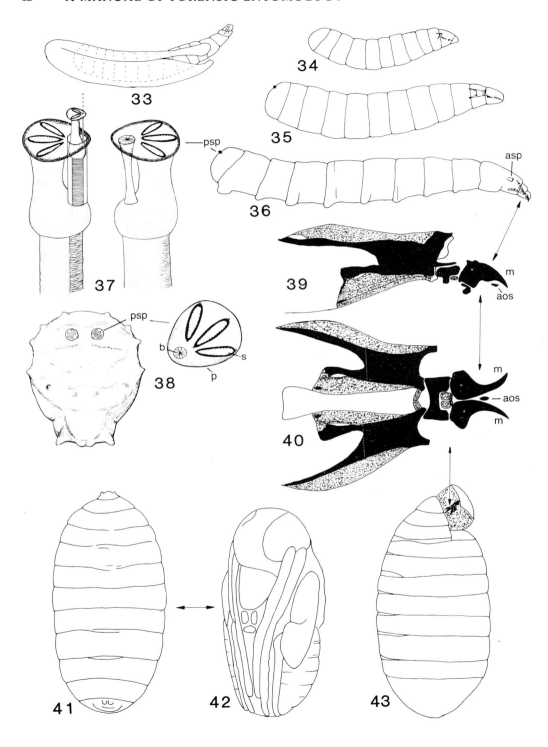

so much the better, since the total development time will facilitate calculation of time of death. Rearing in this way may also reveal unexpected situations such as two different exposures of the cadaver as reported by Nuorteva (1974, see Fig. 48 and Case history 18, p. 66). Where larvae have not been collected on site but have been subsequently collected in the mortuary or during post mortem examination, it is important to establish how long the body has been kept and in what temperature conditions, as this may otherwise have a misleading effect on the apparent rate of development. It is important also to check whether insecticides have been used in the mortuary as this would foil any attempt at rearing the larvae. Where possible, samples should always be collected from the body on site.

Larvae should be reared in moist conditions in containers open to the air, i.e. glass tubes or jars with gauze tops. One larva may survive to pupation in a corked or stoppered tube, but several will die of asphyxiation by their own production of CO_2. Closed containers also encourage mould. The main cause of mortality in open containers is desiccation and care must be taken to keep the larvae and their food source moist by dropping in clean water from a pipette. A few experiments with different containers will soon give one the experience required. Clean empty tins provide a suitable temporary container for a few larvae if not too much food is put in at a time. Nuorteva et al. (1967) used plastic cans with a lid of wire gauze (mesh 0.75 mm, Fig. 25). In the upper part of the can a horizontal exit funnel permits the flies to emerge into a tube where they are collected. The cans were filled with sand and moss to the lower edge of the funnel, and pieces of liver in small vials were used as food. The containers were irrigated to keep the sand and moss moist. Care should be taken not to introduce other invertebrates (not related to the case) with the moss. Only one type of larva should be reared in each container and care must be taken even with these if they are of certain Diptera species which become predaceous on each other in later instars (e.g. *Hydrotaea, Ophyra*).

Any living aquatic insects or larvae collected from a corpse submerged in water could be placed in an aquarium for rearing and/or samples killed and preserved in alcohol for further study. A guide to the literature on rearing insects and mites is provided by Wong (1972).

Specimens for identification should then be selected, e.g. fully grown larvae and adults first. Adults in alcohol can be examined, in a sufficient quantity of the preservative to cover the insect, by manoeuvring them with fine forceps and dissecting needles, or blunt 'seekers'. It may be necessary to impale an insect temporarily to hold it at the correct angle for examination of a particular character (see below). The posterior spiracles of dead but unprocessed fly larvae can be examined by pushing them head (pointed end) first into a watch glass or dish of fine sand, wet or dry. Some diagnostic features (i.e. distance separating the posterior spiracles, the shape of the slits, and the structure of the anterior spiracles) can often be seen by this method, but distinguishing whether or not the spiracular peritreme is closed, and the fine details of the posterior spiracular slits may be difficult until the larva is cleared as described in the section on Equipment for Collection

---

Figs 33–43 Structural details of the immature stages of blowflies. 33, egg with hatching larva; 34, first instar larva; 35, second instar larva; 36, third instar larva (asp = anterior spiracle, psp = posterior spiracle); 37, moulting process of posterior spiracles from second to third instar (after Keilin, 1944; see also Fig. 44); 38, posterior spiracles (psp) with enlarged detail of single spiracle (p = peritreme, b = button, s = slit); 39, mouthparts (cephalopharyngeal skeleton) in side view (m = mandible, aos = accessory oral sclerite); 40, mouthparts (cephalopharyngeal skeleton) in ventral view with lower arrow indicating location inside empty puparium; 41, puparium formed from skin of third instar larva; 42, pupa from inside puparium; 43, empty puparium indicating location of mouthparts of third instar larva.

and Preservation (p. 37). Details of the mouthparts are rarely visible until the larva is cleared.

The stage or instar can be quickly determined for the blowflies and related Diptera by an examination of the spiracular slits to see if it is first instar (one or no obvious spiracular opening), second instar (two slits in each spiracle), or third instar (three slits in each spiracle; Figs 37, 38). If it is a third instar larva it can be identified through the keys to third instar larvae, starting with the key to families. Second or third stage larvae can be compared with illustrations of the mouthparts and spiracles of species it is likely to be (more easily inferred if third instars are also present). Sometimes one may be lucky enough to find larvae at the stage of moult between two instars when both sets of mouthparts and spiracles are visible (Fig. 44). This considerably refines an estimate of the age of the larva. The mature larva has a full crop when it leaves the corpse but this food is gradually absorbed in the non-feeding prepupal stage. The degree of emptiness of the prepupal crop can be used as a further refinement in establishing post-mortem interval, even from photographs (Greenberg, 1985). See also Wolfe (1954).

If intact puparia or pupae are available, some of these should be kept alive in a moist (not wet) container as the subsequent adults will be much easier to identify to species. If necessary, puparia may be separated from the pupation medium by water flotation (Basden, 1947) or forced air (Bailey, 1970).

On emergence adults should be left in their containers for about a day to harden off, inflate their wings and to develop their full pigmentation. This ensures that all the required taxonomic characters are visible. Then they should be killed and examined under the microscope.

The most convenient method for microscope examination is to impale the specimen on an entomological pin either horizontally beneath the wingbases or vertically to one side of the mid-line of the thorax, and to push the pin into a piece of plastazote, 'Plasticine' or the cork from a tube in varying positions. This is often a quicker and more convenient way of handling a specimen than manoeuvring specimens freely floating in alcohol, though they too can be temporarily impaled on a pin or dissecting needle for viewing under a microscope.

If there are enough puparia to spare, one or two should be dissected to see if the pupa is developed inside (Fig. 42). It takes some time for the pupa to form and since larvae are in the prepupal stage for some while (Table 5) the result of this examination may reduce the possible age range. Also the third instar larval mouth-hooks can be retrieved from the cap of the puparium by floating in dilute KOH and can then be processed like a larva as described earlier. These can also be retrieved from empty puparia, provided the part of the puparial 'cap' containing them is still attached or available among the collected puparial debris. Puparia that do not produce adults should be kept as they may contain parasites which will emerge later. (See parasitic Hymenoptera, p. 156.)

The age of adult flies can be estimated by an examination of the fat body and ovaries (Detinova, 1962; Anderson, 1963; Suenaga, 1969). However, this is a difficult procedure and only worth while if it is *certain* that a particular fly has developed in a corpse and that only the adult can provide the evidence required, e.g. from a body indoors at a time when there are no other blowflies outside due to season, low temperatures, etc. These methods can also be applied to other insects (Detinova, 1968). (See Tyndale-Biscoe, 1974.)

Insects other than Diptera (to which most of the foregoing comments apply) are not likely to be of primary importance in the general run of forensic investigations and keys

Fig. 44 Larva of the common bluebottle, *Calliphora vicina*, at the stage of second (2) to third (3) instar moult, showing mouthparts (above) and posterior spiracles (below) of both instars. Such a larva would enable a more precise estimate of age to to be made than a larva showing only the features of one instar (above, lateral view; below, dorsal view). (Photo Peter York; compare with Figs 37–43.)

METHODS AND TECHNIQUES 45

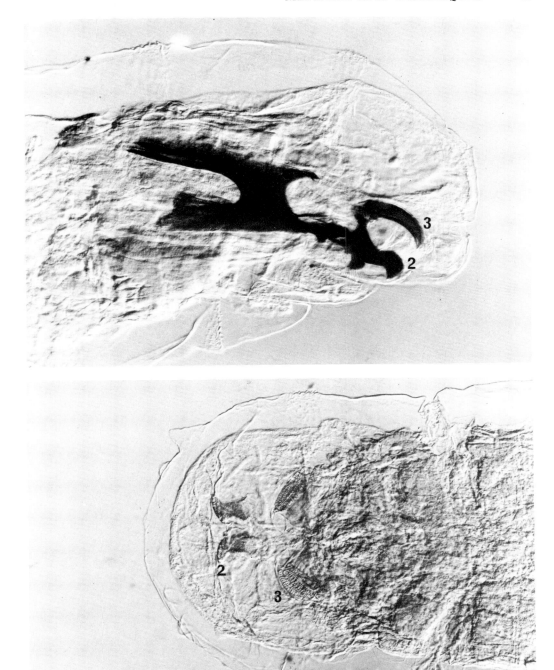

Table 5 Life histories of 11 species of blowflies (Calliphoridae) and flesh-flies (Sarcophagidae) reared at 22°C (71.6°F) and 50% R.H. (After Kamal, 1958)

	No. of Generations	Egg (Hrs)	Duration (Range)					
			First Instar (Hrs)	Second Instar (Hrs)	Third Instar (Hrs)	Prepupa (Hrs)	Pupa (Days)	Total Immature (Days)
Sarcophaga cooleyi	29		24 (18–42)	18 (9–40)	48 (24–84)	96 (42–144)	9 (6–14)	16 (14–18)
Sarcophaga shermani	28		22 (16–38)	16 (8–24)	48 (24–72)	104 (30–168)	8 (6–12)	14 (12–16)
Sarcophaga bullata	18		26 (24–36)	18 (14–24)	54 (30–72)	112 (54–192)	12 (11–17)	17 (16–20)
Phormia regina	23	16 (10–22)	18 (11–32)	11 (8–22)	36 (18–54)	84 (40–168)	6 (4– 9)	11 (10–12)
Protophormia terraenovae	27	15 (12–23)	17 (12–30)	11 (9–20)	34 (20–60)	80 (38–160)	6 (4–10)	11 (10–13)
Lucilia sericata	29	18 (12–38)	20 (12–28)	12 (9–26)	40 (24–72)	90 (48–192)	7 (5–11)	12 (12–15)
Eucalliphora lilaea	27	22 (14–30)	22 (14–36)	14 (9–24)	36 (18–84)	92 (54–204)	6 (4–11)	13 (12–16)
Cynomyopsis cadaverina	17	19 (14–24)	20 (16–30)	16 (12–24)	72 (36–96)	96 (60–168)	9 (7–12)	18 (17–19)
Calliphora vomitoria	5	26 (23–29)	24 (20–38)	48 (43–54)	60 (48–96)	360 (240–504)	14 (11–18)	23 (21–27)
Calliphora vicina	5	24 (20–28)	24 (18–34)	20 (16–28)	48 (30–68)	128 (72–290)	11 (9–15)	18 (14–25)
Calliphora terraenovae	4	25 (18–30)	28 (26–36)	22 (18–28)	44 (30–58)	144 (336–432)	12 (10–17)	20 (19–23)

Table 6 The development of body length (in millimetres) of some fly species during their metamorphosis*, (L = larva, P = puparium, A = adult fly).

Day	Species				
	Musca domestica	Calliphora vomitoria	Lucilia caesar	Sarcophaga carnaria	Piophila nigriceps
2	L 2	L 3–4	L 2	L 3–4	L 1
3	L 2–3	L 5–6	L 2–3	L 5–6	L 2–3
4	L 4–5	L 7–8	L 3–4	L 7–9	L 4–5
5	L 6–7	L 10–12	L 5–6	L 10–12	L 5–6
6	L 7–8	L 13–14	L 7–8	L 13–14	pupariation
7	L 8	pupariation	L 8–9	L 15–16	P 3–4
8	pupariation	P 9–10	pupariation	L 16–18	P 3–4
9	P 5–6	P 9–10	P 6–7	L 19–20	P 3–4
10	P 5–6	P 9–10	P 6–7	pupariation	P 3–4
11	P 5–6	P 9–10	P 6–7	P 10–12	P 3–4
12	P 5–6	P 9–10	P 6–7	P 10–12	A 4–5
13	P 5–6	P 9–10	P 6–7	P 10–12	
14	A 7–8	A 12–13	A 7–9	P 10–12	
15				P 10–12	
16				P 10–12	
17				P 10–12	
18				A 16–18	

*Modified from Schranz: in B. Müller, 1975

to species for families of non-Diptera are not provided in the present work. However, it should be possible to identify the most frequently encountered groups by a careful study of the text and illustrations given in the taxonomic sections of the book.

An elementary knowledge of entomology is assumed, but technical terms have been kept to a minimum. In using the identification keys, both halves of each couplet should be studied before proceeding as only one of the two alternatives may be illustrated. Technical terms are usually explained and illustrated on first mention, but the glossary may also be consulted. Size is not illustrated in the figures, but is given at least for the adults in the taxonomic part of the text (see also Table 6). Illustrations on a block of figures are not usually to the same scale. Illustrations of larvae on a block usually show a whole final instar larva and greatly enlarged details of their structure. Some degree of variation must always be allowed for in biological material and this should be borne in mind when comparing illustrations. Pigmentation patterns of the sclerites of the mouthparts (cephalopharyngeal skeleton) are variable within a species and possibly geographically. The majority of the structural details are from British or European material. With practice one soon gains sufficient experience to use keys and illustrations with some confidence.

Interpretation of data for forensic use

It should always be borne in mind that what we are ageing with entomological evidence from cadavers is usually insects. The body may have been subjected to changes of environment and physical conditions such that the age of the insects and the time of death of the corpse do not appear to coincide. These differences can be brought about if the corpse has been partly or wholly concealed with wrappings or hidden and transported in, say, the boot of a car, or by keeping it indoors for a while and then dumping it outdoors. Simply moving a body from one location to another can complicate the reconstruction of the sequence of events since death (see below).

The interpretation of the entomological data from cadavers will now be dealt with in order of increasing complexity with which cases can occur. Examples of actual case histories are given in a separate section (p. 56) and are cited here by number where relevant to the particular interpretation being made.

If a body is fresh and only blowfly (*Calliphora, Lucilia*) eggs are present in the natural orifices or wounds, it can be assumed that it has only been on site for a very short time (1–2 days) and soon after death. Blowflies do not normally lay eggs at night and, in normal open conditions of exposure from late spring to early autumn, can arrive on a body within minutes and oviposition can take place at once (optimum oviposition temperatures are 15–25°C; see Case history 7, p. 58). If the body is not fresh and only eggs or young larvae of blowflies are present then the body has clearly been stored somewhere where flies had no access. Maggots of different sizes may give other clues. For example, partly covering a body may have inhibited oviposition (see Case history 8, p. 58). Placing a polythene bag over the head of a corpse to stop maggots dropping out onto the upholstery during transport in the back seat of a vehicle can result in more rapid development (due to a higher temperature inside the bag) and, hence, larger maggots than on other parts of the body, even though the body be clothed. Of course, if the build up of CO_2 killed the maggots in the bag those on the body would catch up or overtake them in size! If intact bodies have been completely covered or stored for a long time or kept in dry conditions or deep frozen, they may no longer be attractive to the first waves of blowflies when exposed, and only insects from later in the faunal succession may be in evidence, e.g. *Ophyra* is found in situations of this sort and when bodies have been kept indoors permanently (see Case history 9, p. 59). Entomological evidence can be of particular value here because the good state of preservation of the corpse, especially the internal organs, can be very misleading to the pathologist faced with estimating the time of death.

48 A MANUAL OF FORENSIC ENTOMOLOGY

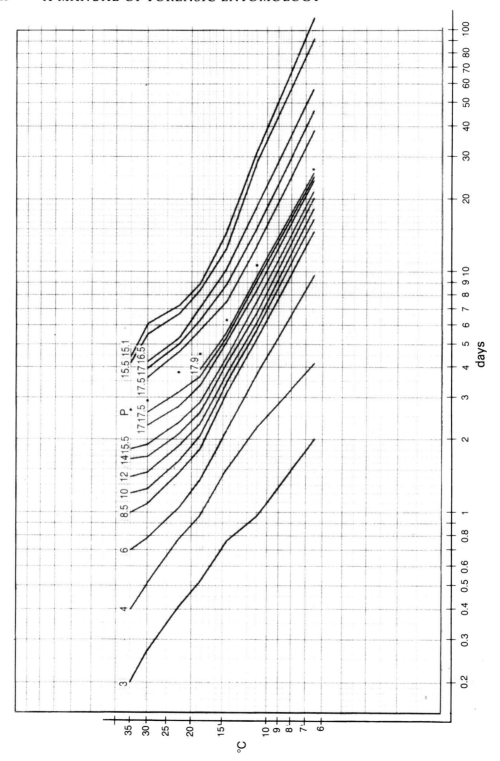

Beetles have been of importance in cases where bodies have been on site for only a short while and maggots have not had access, e.g. new-born babies shallowly buried in, say, a shoe box (see Case history 4, p. 56).

Usually the sort of case in which entomologists are involved is where a person has been killed (or died naturally) on site and remained there and the police are anxious to establish time of death. Once the various insects and larvae have been identified to their species and development stage, their normal place in the faunal succession should be checked (Table 1, Figs 3–5). Next, the prevailing temperatures at the site should be ascertained. If these were not recorded at the time the body was found they can be estimated by obtaining temperature records for the probable period of exposure from the meteorological station closest to the site. The micro-climatic conditions prevailing on site should then be cautiously assessed by interpolation from a close study of the conditions there (see Figs 17–19, Tables 3–6). Rates of development for some common carrion Diptera are given in Table 5 and information for species not included there should be sought in the text. Allowance should be made for the actual temperatures involved, the estimate of age increasing for lower temperatures and decreasing for higher temperatures. If *Calliphora vicina* Robineau-Desvoidy is the species involved, the larvae (fully extended – see collection and preservation) should be carefully measured and compared with the length/temperature/age 'Isomegalendiagram' of Reiter (1984, see Fig. 45). Some other species are treated in Table 6 and other information is given under the appropriate genus or species in the taxonomic sections.

Clearly, commonsense must prevail in these estimates and absolute accuracy will not be possible from data assessed in retrospect unless careful records were made of air and corpse body temperatures on site. Careful assessment can give estimates accurate to a day or so and this is often more accurate than is possible with equal confidence by other pathological methods. As can be seen from the many case histories cited, these methods are frequently successful. An even greater degree of accuracy can be obtained if some larvae are collected and reared in laboratory conditions simulating those prevailing on the site (see Case history 11, p. 60). Nuorteva and Leclercq have skilfully practised these refinements, as is demonstrated in their many publications (see references), and have solved cases with considerable subtlety.

A careful study of what is known of the biology and habits of the insects concerned, combined with actual involvement in a few of these 'standard' cases and discussion with specialist entomologists, soon provides the forensic scientist with the necessary skill and experience to estimate time of death with accuracy and confidence. However, it should always be borne in mind that biological systems are rarely predictable with the precision attainable in the experimental sciences. Allowance should always be made for this. The serious consequences frequently resulting, at least in part, from the forensic entomological evidence, demand that the aim should be for any possible error to be on the side of caution.

The greatest care of course must be taken in identification. The Identification keys in this book are intended as an initial guide for non-specialists and where a group requires a

---

Fig. 45. Isomegalendiagram constructed by Reiter (1984) for *Calliphora vicina*. If the temperature is roughly constant the age of the maggot can be read off instantly from its length. Where temperature is variable an age range can be estimated between the points where the measured larval length cuts the graph at the maximum and minimum temperatures recorded. Air temperature records are usually available from local meteorological stations on request. Where possible on site microclimatic temperatures prevailing in the maggot's immediate environment should be established and correlated retrospectively with the air temperature records. Care should be taken that the maggot is fully extended.

lot of taxonomic experience no keys are given. *When in any doubt the advice of a specialist entomologist must be sought since so much depends on correct identification.*

In addition to time of death it is frequently possible, using entomological data, to ascertain whether a body has been killed on site or moved there from another geographical location, and a careful study of the distribution and habits of the insects found should be done with this in mind. Insects of a very local distribution, e.g. sea shore (see Case history 16, p. 65), can be very useful as indicators of locality and not only in work with cadavers. Insects (even fragments) trapped in the radiators or tyre treads of cars or in the mud on footwear can help reconstruct journeys (see Insects and Transport, p. 167). Insects found in commercial products or containers may indicate their country or area of origin (see Cannabis Insects, p. 169).

Experimental work with carrion

The real need for future research from the forensic point of view is to study the faunal succession on intact human corpses in field conditions. The objections, ethical and moral, to this are obvious. However, these considerations aside, the work would hardly be pleasant though there would no doubt be those with sufficient scientific detachment to undertake it. Any forensic scientist sufficiently interested in the entomological aspects of the subject could do valuable work by investigating the insects in any cases likely to yield information in the categories detailed below. Even where the data are not of particular relevance in the case under investigation, there will undoubtedly be future cases where it will be so and the larger the 'data bank' available the easier these investigations will become.

Meanwhile, studies on animal carrion will continue to be of value provided certain worthwhile ecological aspects are studied. Throughout this book problems worthy of special attention have been mentioned where appropriate and particularly desirable areas of investigations are summarised below:

1. Effect of eating, drinking and working habits before death on decomposition rate and/or the successional fauna after death, e.g. diet, alcoholism, drug-addiction, occupational effects (see quotation from Hamlet on page 5 of this book!).

2. Effect of cause of death on the subsequent faunal succession, e.g. poisoning, gassing, drowning.

3. Effect of treatment of corpse after death on the subsequent fauna, e.g. nature of concealment, deep-freezing, burning, immersion in water, etc.

4. Effect of geographic and ecological location of corpse on faunal succession.

5. The all important specific effects of temperature and humidity still require further refinements.

6. Work on the carrion life histories, biology, behaviour and physiology of the cadaver insects themselves, e.g. chemistry and behaviour of attraction, diel periodicity, nutritional requirements, etc. (see Norris, 1965*a*). Having chosen a line of investigation the choice of carrion bait should be made.

Where possible, experimental work should be carried out in the field with whole carcases. The ideal situation is when an animal is known to have died at a site where it has remained undisturbed and its exact time of death is known. The rearrangement of a corpse, or the use of special equipment to facilitate study, always disturbs the natural sequence of events. In many species the fully grown larvae will leave the corpse and travel as much as 231 ft (6.5 m) away from a carcase before seeking a pupation site (Cragg, 1955; Herms, 1907; Norris, 1959; Nuorteva, 1977; Fig. 32). Over a hard surface, such as in a slaughterhouse, blowflies may travel as far as 100 ft (30–31 m) before burrowing for pupation (A. A. Green, 1951). Cragg (1955) has suggested that this dispersal may be an adaptation to escape the attacks of predators. If this is so then any experimental apparatus which prevents the maggots moving away from the carcase could cause interspecific discrimination. Waterhouse (1947) found that the presence of a rim around a tray designed to catch migrating maggots produced great differences in the numbers of *Calliphora* and *Lucilia* emerging from sheep carcases.

Similarly, apparatus which excludes rain or sunlight not only alters the temperature and moisture regime on/in the carcase but could exclude certain species behaviourally, e.g. the heliotropic *Lucilia* species.

The use of butchered meat produces unrealistic results in carrion studies because larvae are deprived of many natural niches which may render them more vulnerable to predation and parasitism. Such meat also desiccates quickly. However, for studies on behavioural aspects of carrion insects 'artificial baits' may be useful.

The different behaviour exhibited by different genera of blowflies should be considered for both the choice of bait and the site in which it is placed (Parish & Cushing, 1938; the series of papers by MacLeod & Donnelly, 1956–63). Blowflies are attracted to a wide range of substances, particularly flesh, ripe fruits, sugars, faeces, urine, etc. but not all of these are conducive to oviposition. Adults mainly obtain their energy from the nectar of flowers. Wardle (1921) found that the following foodstuffs, which are suitable for human consumption, were the most attractive to oviposition and in order of preference they are: raw liver, raw sheep kidney, raw lean mutton, raw lean beef, beef cooked but underdone, and mild cured bacon. He found chilled meat, well-cooked meat, over-ripe fruit, preserved meats (sausages, tinned meats, salt bacon), tripe (as prepared for retail sale), fresh fish, animal fats, cheese and shellfish unsuitable. Perhaps the second list holds a message for us all and confirms the claim of those graffiti writers who assure us that in their choice of a certain foodstuff 50 million flies can't be wrong! Fish was only attacked when far too stale for human food and was the bait chosen with considerable success during 'National Fish Skin Week' when the Blow-Fly Recording Scheme was launched by my former colleague James Dear (1981; though the possible consequences once the 'media' learn of such a scheme should be heeded (Smith & Dear, 1978)). Plants with a carrion-like smell may also serve as baits (Austen, 1896; Smith, 1956) as can 'honeydew' secreted by aphids (Nuorteva 1961a – see also Insects and Transport, p. 167).

Wardle (*op. cit.*) thus established that blowflies prefer to oviposit on substances which contain animal proteins, especially albumins and globulins. The main stimulus to oviposition seems to be exuding juices and muscle plasma. Blood alone, fresh or putrid, seems unattractive as observed by Lodge (1916), though blood-stained clothing may be attractive if stale and wet (Nuorteva, 1974; Case history 18, p. 66). Lodge also noted that blowflies are more strongly attracted by putrefying substances than by fresh ones, especially when some maggots are already present and the putrefaction is thus enhanced by their digestive juices.

The site chosen may favour some blowflies more than others. Bluebottles (*Calliphora*) prefer shade while greenbottles (*Lucilia*) prefer sunlight. Wardle (1921) cites the most convincing and poignant proof of this that I have met in the literature. He writes: 'This difference was strikingly observed during the hot August sunshine of 1916 in the trenches outside Delville Wood on the Somme, where blowflies were abnormally abundant. In the shade afforded by the deep portion of the trench, round the traverses, any moist patch on the chalk wall would be hidden by a dense, indigo-coloured cluster of *Calliphora*, large as a soup plate, unredeemed by the metallic green of *Lucilia*, whereas, where the trench was shallow or blown in, the green shimmer of *Lucilia* was everywhere.'

The method of killing animals used in experimental studies can also affect the rate of development, e.g. gases can combine with haemoglobin and affect decomposition rates, although of course comparative studies could be useful here if a gas or poison has been used in a murder case. If corpses are split open they desiccate more quickly, which can also affect results, though here again precise data comparing the attractiveness of natural orifices and artificial ones (wounds) are not available and this is worthy of investigation.

The effect of burning on the subsequent fauna of corpses has not been fully investigated and forensic cases involving burning are not infrequent, e.g. ritual suicide, infanticide, arson to cover up a murder inside a building.

Bodies are frequently transported after death and dumped away from the murder site. Such corpses may be temporarily concealed in a freezer, the boot of a car or a trunk. Blowflies and other insects may thus be prevented from ovipositing on the fresh body and when 'dumped' it may no longer be attractive to the first waves of insects, e.g. blowflies. Comparative studies on carcases subjected to such treatment with similar control carcases are required. Payne (1965) found that meat (baby pigs) could be attractive to blowflies within minutes of removal from the freezer!

There are many cases of mummified bodies or corpses concealed under floorboards in buildings where only later members of the faunal sequence have gained access, such as the fly *Ophyra* or various 'stored products' type beetles and moths. Carcases can be placed in such situations for experimental investigation.

Larger carcases may need to be protected from vertebrate predators by a strong cage or container constructed of wire mesh and securely anchored. Smaller carcases can be placed in traps. A convenient trap (Fig. 24) used by my colleague Dr R. P. Lane (1975) is constructed from a plastic dome of the type frequently used by horticulturalists. Flies are removed from the trap by sliding a piece of cardboard between the base of the dome and the bait. Ammonia (80%) vapour is then introduced from an inverted wash-bottle to 'knock down' the flies. Inverting the wash-bottle ensures that no liquid escapes into the trap and the card ensures that the bait does not become contaminated with ammonia. Sample collection takes only a minute and very few insects escape. The height at which traps are set should be varied. Roberts (1933) found that the numbers of necrophagous insects caught decreased with height while the numbers of their parasites caught increased. Ground traps (Fig. 23) can also be baited with carrion in order to catch the ground fauna, such as beetles. More elaborate methods of trapping insects, some relevant to carrion studies, are described by Cogan & Smith (1974), Southwood (1978) and Stubbs & Chandler (1979).

The marking of adults permits the study of their dispersal and local migration. Norris (1957) conceived the brilliant idea of letting blowflies emerge in sand mixed with fluorescent dust, so that as they emerged and the ptilinum (see next section) deflated it automatically trapped particles of dust in its folds. The flies could then be released and identified on recapture under ultra-violet light (see also MacLeod & Donnelly, 1957a).

Once specimens are trapped or collected the same procedures can be followed as described in the sections on forensic procedure on site and in the laboratory (p. 39 and p. 41 respectively). General treatments of carrion ecology, especially the theoretical aspects are given by Elton (1966), Putman (1983) and Varley et al. (1975). Model studies on how the carrion fauna should be investigated are provided by Akopyan (1953), Bornemissza (1957), Braack (1981), M. Coe (1978), Fuller (1934), Hennig (1950), Lundt (1964), MacLeod & Donnelly (1958–63), Nabaglo (1973), Nuorteva (1977), Nuorteva et al. (1967, 1974), O'Flynn (1983), the series of publications by Payne et al. (1965–72) and Reed (1958). These and other studies are referred to throughout this book and provide information from the major zoogeographic regions and climatic zones.

Life histories

The following brief and simplified account of insect life histories is given for the benefit of readers without entomological training. Fuller accounts will be found in the text-books cited at the end of this section. Additional information is also given at appropriate places throughout this book, especially under individual orders and families.

Insects that undergo an incomplete metamorphosis hatch from the egg resembling miniature versions of the adult, except for the absence of wings (in some groups wings are also absent in the adults, e.g., lice, springtails). The wings first appear as small pads, but are non-functional until the final moult. This immature form is frequently referred to as a nymph and attains the adult stage by a succession of moults. Examples of this type of insect of forensic interest are the cockroaches (Blattodea), earwigs (Dermaptera), termites (Isoptera), bugs (Hemiptera) and lice (Phthiraptera). The precise identification of the nymphal stages can be very difficult and if the species cannot be inferred from the presence of associated adults the advice of a specialist should be sought.

Those insects that have a complete metamorphosis hatch from the egg as a larva which is quite different in appearance from the adult, e.g. the maggot of a fly, the caterpillar of a moth, the larva of a beetle, wasp or flea. After a series of moults the larva enters a resting stage called the pupa. Sometimes the larva will spin a protective cocoon inside which it pupates. Examples of this type of insect of forensic interest are the flies (Diptera), beetles (Coleoptera), ants, bees and wasps (Hymenoptera) and caddis-flies (Trichoptera).

In the more specialised flies (Diptera, Cyclorrhapha) there are three larval stages or instars (distinguished by the presence of one, two or three slits in the posterior spiracles (Figs 34–38)) and the third stage larva forms a puparium from its cast skin, inside which the pupa develops and out of which the adult eventually emerges (Figs 33–43). When feeding is completed the larvae leave the corpse and search for a site suitable for pupariation and pupation. Prior to forming the puparium the larvae become less active and contract their bodies into a shorter thicker form, or prepupa. At this stage changes in environmental conditions, such as lower temperatures or humidity or less light or oxygen, may induce diapause (suspended animation in which development is arrested – see Glossary, p. 174) (Cousin, 1932; Cragg & Cole, 1952; A. C. Evans, 1935; Mellanby, 1938; Ohtaki, 1966; Vinagradova & Zinovjeva, 1972; Whiting, 1914; Zakharova, 1966). Autumn temperatures of about 7°C induce diapause in blowflies, but sarcophagids enter diapause at higher temperatures and by a complex interaction of temperature and daylight (Zakharova, 1966). Some species of *Calliphora* and *Phormia* may have no clear diapause, but low temperatures will arrest their development (A. A. Green, 1951; Tamarina, 1958, 1967). A knowledge of the types of diapause in blowflies may be important in forensic cases in establishing whether death has occurred in late autumn or early spring. See also Fraser & Smith (1963).

The prepupae become puparia by a process of hardening and tanning of the larval skin from white to dark brown. This can be an important stage in forensic cases as puparia can be aged from their colour (see Procedure on Site, p. 39). The pupa is formed within the puparium. The puparial stage of most sarcosaprophagous flies takes about half of the total life cycle from egg to adult.

Emergence of the adult is effected by inflation (with fluid) of the ptilinal sac, on the head, which ruptures the puparial cap. After emergence adult flies use the expanding and contracting movements of the ptilinal sac to help them move through the soil. Immediately after emergence the flies are small and soft, but as the air sacs fill and the wings expand the fly attains normal size. This takes about 30 minutes and about one and a half hours after that it is fully coloured (Fraenkel, 1935).

Non-insect groups of Arthropods included in this book have a life-history resembling the incomplete metamorphosis type of insects, e.g. mites and spiders (Arachnida), prawns, shrimps, crabs, water-lice, etc. (Crustacea).

Identification

The taxonomic literature on insects is vast and clearly the scientist wishing to conduct original research in the forensic application of entomology will need access to a good library and to form his own collection of books and papers. The Identification Keys and

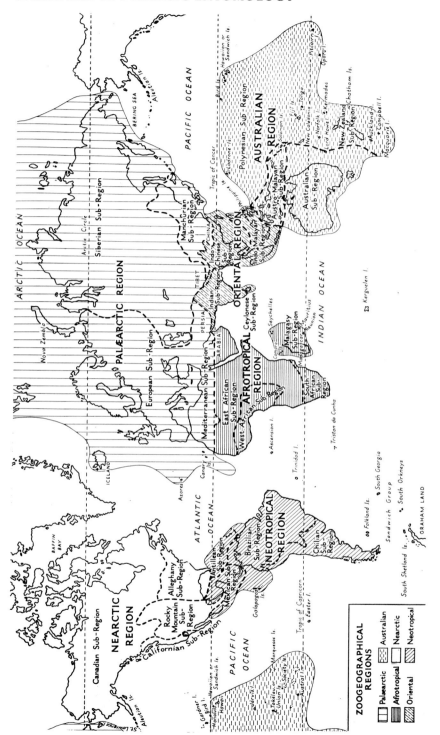

Fig. 46. Zoogeographical regions, from Bartholemew, Eagle Clark and Grimshaw's Atlas of Zoogeography (Bartholemew's Physical Atlas, Vol. 5, 1911).

illustrations in the present work should facilitate identification of most of the British insects encountered in the normal routine of forensic enquiries involving estimates of time of death based on cadaver fauna and other aspects.

As far as possible comprehensive references are given which will facilitate identification of less-frequently encountered species, at least for Britain and as far as possible for the Palaearctic region. (Often British works will cite the literature of other regions in their references.) Workers in areas other than the UK wishing to assess their insect fauna should consult the bibliographies by Kerrich *et al.* (1978) for works facilitating identification to species level for Britain and northern Europe, and Hollis (1980) for works on world insects enabling identification to generic level. These bibliographies also cite the standard catalogues which list all the known species occurring in a particular region (see Fig. 46 for map of zoogeographic regions). The orders and families likely to be involved in forensic work will be, in the main, the same throughout the world, but differences are to be expected at generic and specific levels, except for those with a cosmopolitan distribution (e.g. *Calliphora, Lucilia, Musca domestica* L.). A study of the available literature (see Bibliography, p. 179) on carrion fauna for the region under consideration will soon indicate the principal species to be involved in that line of enquiry.

Clearly, accuracy of identification is vital and *the help of expert entomologists should always be sought in cases of difficulty or doubt.* The expert will know the group as a whole and not just within the narrow confines of normal forensic application, and this can be extremely helpful in unusual and complicated field situations (see Case histories, especially comments to Case history 3, p. 56).

It is recommended that any forensic workers intending to use entomological methods on a regular basis should thoroughly familiarize themselves with the blowflies and other carrion insects occurring in their particular area. Particular attention should be given to the identification of the immature stages since, for the most part, these are not covered comprehensively by the taxonomic literature and should be checked by rearing adults.

Accurate observation and careful documentation will help to establish the collection of entomological data as a standard routine in forensic investigation. The more cases that are investigated the more precise will the method become.

For the benefit of readers who lack entomological training brief accounts of the life histories of insects involved in forensic work and much incidental and relevant detail is given throughout the book. Further information may be gleaned from the general text-books of entomology and other comprehensive works such as Chapman (1982), Chinnery (1973), Imms (1971, 1977), Grassé (1949–51), Kükenthal (1968), I. M. Mackerras (1970), Oldroyd (1964), Smith (1973*b*) and Wigglesworth (1964, 1984). General sources of entomological information are given by Gilbert & Hamilton (1983).

Case histories

The case histories cited here illustrate specific examples of the forensic application of entomological methods. Where appropriate, comments are made on each case and each is also numbered to facilitate cross referencing from other parts of the book. The earlier accounts are unavoidably sketchy due to the novelty of the method at the time. Many cases are reported only in restricted publications and are therefore not accessible for analysis.

1. Dr Bergeret's case (1850)

This appears to be the first published account (Yovanovitch, 1888) of the application of entomology to legal medicine.

Dr Bergeret d'Arbois of Jura performed an autopsy on the body of a child discovered by a plasterer while repairing a mantelpiece. He found that the flesh-fly, *Sarcophaga carnaria*, had deposited larvae in 1848 and mites had laid eggs on the dried corpse in 1849, and concluded that judicial suspicion should fall on the occupiers of the house in 1848.

2. The Ruxton case (1937)

This classic case (Glaister & Brash, 1937) began during the afternoon of 29 September 1935, when the police were informed that dismembered human remains had been found in a river near Edinburgh. Two bodies were re-assembled from the remains and proved to be a Mrs Ruxton and her children's nurse, Mary Rogerson. The date on which the remains were deposited was established by the presence of third instar larvae of *Calliphora vicina* Robineau-Desvoidy (as *erythrocephala*) of an age estimated at 12–14 days by Dr A. G. Mearns. This estimate agreed with and corroborated other evidence and led to the conviction of Dr Ruxton, although he did not confess.

3. The Lydney case (1964)

This case is of interest in that the prosecution's pathologist, Professor Keith Simpson, successfully used blowfly maggots to establish time of death, in spite of an attempt by the defence counsel to refute this evidence by calling an entomologist as an expert witness. The case hinged on whether the testimony of three witnesses, who claimed that they had seen the victim after the supposed date of death, outweighed the evidence of the prosecution which finally depended on the age of *Calliphora vicina* Robineau-Desvoidy (= *erythrocephala*) maggots found on the corpse (in June). Simpson aged the maggots as 'at least nine or ten days old, but probably not more than twelve.' In the witness box the entomologist (A. W. McKenny Hughes) concurred.

Simpson's (1980) graphic account of this case illustrates the pressures one may be placed under when offering entomological evidence and the importance of a knowledge of the fine detail of blowfly biology in supporting this evidence.

As an example of the value of entomological evidence Professor Simpson's estimate of the age of the third instar *C. vicina* larvae gave a time of death of 'at least nine or ten days, but probably not more than 12', contrasting sharply with the police estimate, from the state of disintegration of the body (in June), of 6–8 weeks!

4. The baby in the box (Easton, 1944; Easton & Smith, 1970)

On 15 March 1944, in freezing conditions, the body of a new-born child, with placenta attached, was found wrapped in a blanket and newspaper in a cardboard box placed in a slit trench. The floor of the trench was deeply covered with the previous year's beech leaves. From its well-preserved condition, the pathologist who examined the body considered it to have been abandoned but a few hours. However, the presence in the wrappings and about the placenta of approximately thirty examples of the Staphylinid beetle *Anthobium atrocephalum* Gyllenhal (Fig. 332) suggested a much longer period. Moreover, the weather had been continuously very cold to freezing during the previous two weeks and under such conditions the beetles would have required time to migrate in such numbers from hibernation among the beech leaves. The mother's subsequent confession revealed a time-factor of nine days.

5. *Kathleen McClung (Easton & Smith, 1970)*

Kathleen McClung was murdered in Guildford during the night of 20–21 June 1969. Post-mortem examination was performed by Dr Keith Mant on June 24 and two days later Dr A. M. Easton received from him two tubes of maggots. The first had been obtained from the mouth. They were small and dead and their identification proved difficult. The second tube contained numerous live maggots obtained from the polythene bag which had been placed around the head of the deceased before removal [to the mortuary]. They were second-instar larvae, the determination of which on morphological characters was again doubtful. However, some of them were reared on a diet of raw beef and a total of eight puparia were obtained between 4 and 8 July. All hatched between 18 and 23 July, the resultant flies proving to be the common *Calliphora vicina* Robineau-Desvoidy. Estimates made from consideration of the dates of pupation suggested that oviposition had occurred between 21 and 24 July – a conclusion closely in keeping with the known facts.

6. *The child behind the stove (Leclercq, 1969)*

On 21 May 1947 the police found the body of a child behind a stove in a farm at St Hubert (Belgian Ardennes). The body was wrapped in a linen cloth in which, at the time of the discovery, there were numerous larvae of *Calliphora vicina* Robineau-Desvoidy in the final stages of their growth; in addition there was a dead female of *C. vicina* (a specimen which had died during hibernation after laying eggs), a quite recent pupa of the same species and some pupae of Phoridae. The *Calliphora* larvae had nibbled at the face of the child, causing the disappearance of the eyes and skin; they had penetrated into the frontal sinuses and from here had devoured the brain. The neck and the upper parts of the arms, as well as the viscera, were also severely damaged.

The larvae of *Calliphora* produced all their pupae between 21 May and the evening of 22 May; the adult flies appeared from 2 June onwards, perhaps ten days after pupation. J. & M. Leclercq (1948) had earlier reared numerous specimens of *C. vicina* and had been able to determine with the greatest precision that during the spring under the thermal conditions of a lightly warmed interior room, the temperature of which had never exceeded 20°C, and under good nutritional conditions on fatty cheese, the development of a batch of eggs of *Calliphora* required 19–20 days from the day when the eggs were laid to the formation of the first pupae.

One can suppose that the larvae found on this corpse underwent development under comparable conditions. In effect: (a) they developed during the spring season; (b) the corpse had been left behind the stove which was sometimes lit and, consequently, the temperature conditions should have been appreciably like those of a lightly warmed interior room, all the more so because the month of May 1947 was relatively warm. The Leclercqs had excluded the hypothesis that the larval development had been accelerated by temperatures higher than those of their experiments, because the stove had not been alight all the time and it was evidently sheltered from the rather exceptional rises of temperature which occurred on some of the days of May 1947.

They therefore agreed that there was a strong presumption that the eggs had been laid about 20 days before 21 May, perhaps about 1 May 1947. Moreover, the eggs laid by *Calliphora* must have been laid on the corpse a short time after it had been abandoned. In fact:

1. *Calliphora vicina* is common throughout the year, present in rural houses and passes the winter in the adult stage.
2. The females very readily detect the odour of flesh that is beginning to decompose. As it was a case of a corpse abandoned in the open air at a time of the year favourable to rapid putrefaction, only a few days would have been needed before the first blowfly arrived to lay eggs.
3. It is known that *Calliphora* belongs to the first wave of necrophagous species which colonize a corpse in the open air.
4. It was the first generation of *Calliphora* which had been able to develop on the corpse. Every earlier generation would have left traces such as empty puparial cases under the corpse or in the cloth covering it.

Therefore they formed the hypothesis that the corpse was placed where it was during the last week of April, a little after the murder of the child. The judicial inquiry took its course and the culprit was arrested; his declarations and confessions completely confirmed the Leclercq's conclusions.

7. *The erroneously condemned Hungarian ferry skipper (Nuorteva, 1977)*

A ferry skipper had been condemned to life imprisonment for the murder of a postmaster, whose knifed body had been discovered one evening (in the month of September) on the ferry. The ferry skipper had arrived at work at 6 p.m. on that day and the body of the murdered postmaster had been found some hours later. The autopsy was performed the next day at 4 p.m. Masses of yellowish fly eggs and numerous newly hatched larvae of 1 to 2 mm in length were present, and the finding was recorded in the autopsy report. No attention was paid to this observation at the trial, however. On assumed evidence, the ferry skipper was condemned to life imprisonment in spite of his swearing that he was innocent. Eight years later the case was reopened. At the new trial, Dr Mihályi pointed out that no sarcophagous flies are active in Hungary at 6 p.m. in the month of September. He also recalled some of his experiments indicating that, at a temperature of 26°C, the yellowish eggs of *Lucilia caesar* (L.) hatch after 13 hours, those of *L. sericata* (Meigen) after 10–11 hours, and those of *Phormia terraenovae* Robineau-Desvoidy 14–16 hours after oviposition. These data, applied to the case of the ferry skipper, led to the conclusion that it was not possible that the eggs could have hatched if they had been laid during the day of the autopsy and that they must have been laid during the previous day *before* 6 p.m. Dr Mihályi's data on oviposition were verified and, on the basis of this and other evidence, the ferry skipper was released from prison.

8. *The two murdered hitchhiking girls (Nuorteva, 1977)*

On 21 August 1971, at 4 p.m., the corpses of two murdered girls who had been hitchhiking were discovered in a sandpit near the town of Hyvinkää, in southern Finland. The corpses were partially covered by a polyethylene sheet. A cluster of fly eggs was collected from the hair of one of the girls; a fly larva between 4.5 and 5 mm long was also present in one eye. Four days later, examination of the refrigerated bodies revealed four larvae 5–6 mm long in the eyes of the same girl, and five larvae 2.5–3.5 mm long in the eyes of the other girl.

An attempt was made to rear all the eggs and larvae to adult flies. The development of the eggs into larvae 4.5–5 mm long (i.e., to the length of the fly larva observed on 21 August required one and a half days. Further rearing was only partially successful and a single adult fly of the species *Calliphora vicina* was obtained. Flies of the same species were also obtained from a liver growing-medium placed in a glass container on 28 August 1973 in the place where the dead girls had been found.

Since it had taken one and a half days to obtain a larva 4.5–5 mm long experimentally (the same length of the larva detected in one of the dead girls), it was concluded that the bodies had been in the locality where they were found from 19 August, namely for about two days after the time of death. The suspected murderer, however, had an alibi for 19 August and for the following days as well. During the trial the question was raised whether it was possible that the dead girls could have been in the place where they were found on 14 August, as suggested by the police investigation. The answer to this question was that, considering the daily temperature from 14 to 19 August (well above 16°C during each day), a large number of big fly larvae should have been found in the corpses. Since this was not the case, one had to draw the conclusions that either the corpses of the girls were not at that place on 14 August or they were completely covered by the polyethylene sheet. From the photograph taken by police immediately after the discovery of the corpses, it seemed possible that the polyethylene sheet had at first covered the girls completely, but later had been partially removed by the wind. The subsequent question was whether fly oviposition might have taken place through possible holes in the sheet. The polyethylene sheet was immediately inspected, but no holes were detected. The results of the police investigation, substantiated by the entomological observations, led to the conviction of the suspected murderer.

9. *The lightly covered corpse in the Forest of Ylöjärvi (Nuorteva, 1977)*

On 13 and 14 July 1970, the Finnish Central Criminal Police submitted nine soil samples collected from a forest in Ylöjärvi (central Finland), where the decayed corpse of a 17-year-old girl had been found on 9 July partly covered by moss and by branches of a rotten tree. Most of the soil samples revealed only non-indicative insects, but the sample collected from the area where the head had been lying contained 187 larvae of blowflies in different stages of development, plus eight scarabaeid beetles of the species *Geotrupes stercorosus* (Scriba), histerid beetles of the species *Hister unicolor* L. and *H. striola* Sahlberg, and staphylinids of the genus *Philonthus*. This soil sample was polluted by fluids from the decayed brain. Another sample of soil from where the pelvis of the corpse had been lying contained 91 fly larvae (all of small size) plus three *Geotrupes stercorosus* beetles and 18 staphylinid beetles of the genus *Atheta*. This sample also contained 17 fly puparia, of which seven were empty. The puparia and some of the larvae were placed in rearing cans.

From 16 to 20 July two specimens of *Muscina assimilis* (Fallén) and some other muscids emerged from the puparia, but not a single blowfly. From the blowfly larvae taken under the head and reared on raw liver in outdoor temperatures, 34 adults of *Phormia terraenovae* Robineau-Desvoidy emerged from 27 July to 4 August, three adults of *Lucilia illustris* (Meigen) emerged from 30 July to 3 August, and seven adults of *Calliphora vomitoria* (L.) emerged from 1 to 3 August.

Since no puparia of blowflies could be identified, but only full-grown larvae were detected, entomological evidence indicated that the corpse could not have been in the area where it was found much longer than a week. In contrast, the existence of mature pupae of *Muscina assimilis* indicated an earlier time of death, but the possibility existed that they belonged to the normal fauna of the locality. Although the entomological observations indicated that the corpse could not have been exposed to fly oviposition for much longer than a week, the advanced brain decay suggested that the death had occurred considerably earlier. Therefore the question arose whether the murder had occurred elsewhere in a flyless environment and the corpse later transferred to the locality where it was found. Information was subsequently received that the corpse was covered by a rather thick layer of moss and that only the head, one breast, and a hand were uncovered. Probably the cover had been quite complete at first, but later had been torn by foxes, dogs, or other sarcophagous vertebrates. This exposure to blowfly oviposition had presumably happened about one week prior to the detection of the corpse.

Later, the brother of the girl confessed that he had accidentally killed his sister with a karate blow to the side of the neck on 4 June 1970. He further stated that in order to hide her death he had carried the body into the woods, arranged the clothes and position of the corpse to simulate a sexual murder, and then covered it with moss and rotten tree branches.

This shows how effectively even a thin cover on the corpse may inhibit fly oviposition. It also shows how important it is to observe closely the type and extent of the coverings on the corpse. Likewise, it indicates that flies of the genus *Muscina* have the peculiar ability to oviposit without immediate contact with the carcase.

10. *The crime of the cleaning woman (Nuorteva, 1977)*

One summer day, a high official of the Finnish Government, on entering his office, accidentally noticed numerous large, white larvae under the carpet at the threshold. He immediately called the cleaning woman into the room and asked her how often the carpet was cleaned. She assured him that the carpet was cleaned every day and that it had been done the previous evening. The official replied that he could not believe that 'bugs' longer than 1 cm could have developed in a single night. The cleaning woman was dismissed because it was assumed that she had told a lie.

Out of mere curiosity a veterinarian working in the same building was requested to inspect the carpet and he wondered whether it was possible that the larvae had really developed by eating the carpet, which was made of a plastic material. Specimens of larvae were collected and sent to Nuorteva for identification. They were identified as migratory larvae (yellowish fat body) or white prepupae of blowflies of the species *Lucilia sericata* (Meigen). It was concluded that the larvae had developed on the carcases of some dead mice or on some forgotten provisions in the government buildings and that they had migrated for pupation during the night into the carpet on the threshold. The cleaning woman was reinstated in her job.

11. *The decayed corpse in a park in Helsinki (Nuorteva, 1977)*

On 8 July 1973, the decayed corpse of a man was found in an isolated corner of a park in the suburban area of Oulunkylä, in Helsinki. The corpse was camouflaged by a few tree branches, but the chest and one hand were quite bare. The skin on these areas was tanned and dry. Thousands of large fly larvae were creeping over the body. In addition, numerous fly puparia were detected in the soil under the head of the corpse. Two tree branches, numerous fly larvae, a puparium and soil samples were sent to Nuorteva for examination. One of the branches, from a rowan tree (*Sorbus aucuparia* L.), had dried leaves 10–12 cm in length. The length of the leaves of a fully-grown rowan tree, as a rule, range from 10 to 16 cm. Fortunately, published information (Nuorteva, 1952) existed on the development of the length of the leaves of the rowan tree in Helsinki during the year 1949. According to these measurements, the growing leaves had reached the mean length of 8.4 cm on 20 May and soon after they had reached full length. The length of the rowan leaves found on the corpse thus indicated that the branches used to cover the body dated from the end of May or the beginning of June.

It was also noted that one of the rowan leaves was inhabited by a colony of moth larvae that had spun the leaflets together. Dr J. Kaisila at the Zoological Museum of Helsinki University determined that these larvae belonged to the species *Hyponomeuta malinellus* L., and Dr Junnikkala at the Department of Physiological Zoology advanced the information that these moths spin the leaflets of the rowan tree together in late May or the beginning of June. Thus, these estimates supported the data based on the degree of leaf growth.

It was further observed that some of the puparia collected from the soil were quite fresh (i.e. they were white prepupae) and intact. The puparia were placed outdoors in a Nuorteva-type rearing can (Fig. 25; Nuorteva *et al.* 1967), under micro climatological conditions corresponding to those where the dead body had been found. From 12 to 22 July, 93 adult blowflies of the species *Phormia terraenovae* Robineau-Desvoidy emerged.

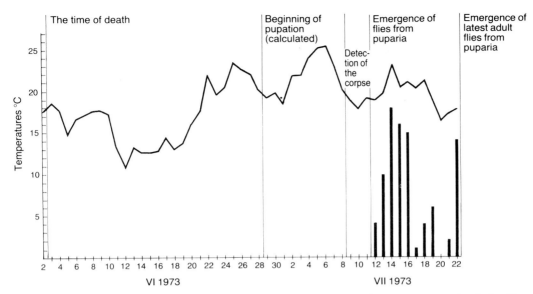

Fig. 47. Case history 11: environmental temperature and emergence of adult flies (*Phormia terraenovae*) from puparia collected from a decayed corpse found in a park in Helsinki. The occurrence of white prepupae as well as the absence of empty puparia made it possible to evaluate the duration of puparial time by rearing under field conditions. Usually the duration of complex development is twice that of the puparial duration, but in this case the low temperatures prior to 28 June had delayed larval development (from Nuorteva, 1977).

The development from white prepupa to adult fly had thus lasted 14 days (from 8 July to 22 July) and the oldest flies emerged on 12 July. Meteorological data showed that the larval development had occurred during a cold period (16.8°C as a mean), whereas it had been considerably warmer (19.4°C as a mean) during the puparial time (Fig. 47). From this evidence it was concluded that fly development on the corpse had started considerably earlier than 15 June, most likely at the beginning of the month. Independent police investigation showed that the victim of the murder had been knifed on 2 June. Thus, the entomological conclusions were correct.

This case is instructive inasmuch as it shows that it is possible to calculate the whole duration of development of flies by field rearing restricted to white prepupae and puparia. The error would have been greater if the estimate had been based on available laboratory evidence on the duration of development for this species (Berndt & Groth, 1971; Kamal, 1958).

12. *The murdered girl hitchhiker in Inkoo (Nuorteva, 1977)*

On 19 August 1971, a police inspector submitted fly samples collected from the badly decayed corpse of a young woman discovered in a small pit in the rural parish of Inkoo, which lies not far from Bromarv (south Finland). The sample consisted of some full-grown blowfly larvae, 21 puparia, and one freshly-emerged adult fly, with still wrinkled wings. The fly was identified as *Phormia terraenovae* Robineau-Desvoidy, and it was determined later, by rearing, that the larvae and puparia also belonged to the same species.

Fortunately, on 4 July 1971, a field experiment investigating the duration of development of blowflies fed on fish bait in localities of different types was in progress. The flies for this experiment had thus developed under the very same meterorological conditions as the flies in the unidentified corpse in Inkoo. For the different fly species in the experiment the minimum duration of development was as follows (at a mean temperature of 16.6°C during the period from 7 July to 8 August):

Lucilia silvarum (Meigen)	28 days
L. illustris (Meigen)	32 days
Muscina assimilis (Fallén)	32 days
Lucilia richardsi Collin	38 days
Cynomyia mortuorum (L.)	38 days
Calliphora erythrocephala (Meigen)	38 days

Phormia terraenovae was not included in Nuorteva's experiments, but it was known from the experiments of Kamal (1958) that the minimum duration of its development is 71% of that of *Calliphora erythrocephala* (= *C. vicina* Robineau-Desvoidy; see Table 5). Under the existing meteorological conditions, the duration of development for *P. terraenovae* should thus have been 26–27 days or probably longer, because the corpse was lying in a pit under a cover that shaded it. Therefore it was estimated that death had occurred more than a month previously. Later it was proved that this estimate was correct. The girl had been murdered on 12 July 1971.

In this case the mercury content of the flies emerging from the rearing of the puparia was determined by activation analysis in collaboration with Dr Erkki Häsänen (Reactor Laboratory in Otaniemi). The mercury content was 0.12–0.15 ppm (fresh weight), which indicated that the flies had developed in unpolluted biological material. Blowflies are able to accumulate mercury to highly elevated levels when they develop in a biological substrate with a mercury content surpassing the natural level (threshold 0.2 ppm; Nuorteva & Häsänen, 1972). Consequently, in this case the conclusion was drawn that the murdered girl had not lived in a mercury-polluted area. Later, when the corpse was identified, it was discovered that she had lived as a student in the city of Turku, which is an area without mercury pollution (Nuorteva, 1971).

This provides the first indication that mercury analyses of this kind may find application in forensic practice. Mercury analyses of the hair of the subject under investigation is advisable, but flies are more sensitive. In the present case the mercury content of the hair was determined; it was 0.30 ppm, which is characteristic of populations that are not exposed to mercury contamination.

13. *The body in the flat (Nuorteva et al. 1967)*

The body of a woman was found on the floor of her flat in Helsinki on 6 September 1964. The body was lying in a place exposed to sunshine through a closed window. Death had obviously occurred on 6 August 1964, because newspapers had accumulated since that day. The cadaver was thus about one month old and considerably decayed. At autopsy, heart failure was found to have been the cause of death. Numerous fly larvae were crawling on the skin and in the orifices of the face. Numerous larvae were also observed on the floor of the flat.

The larvae were collected from the skin surface of the cadaver on 8 September during the autopsy and were transferred to the rearing cage (at low room temperature). The bulk of the larvae were at this time full-grown and actively crawling, and all had reached the full-grown stage on 18 September. The first puparium was detected on the same day. The larvae had eaten only a bit of decayed human liver, but had not touched a piece of cattle liver, which was fresh at first. The first fly imagines emerged on 26–27 September and proved to be *Calliphora vicina* Robineau-Desvoidy (three male, one female) and *Lucilia sericata* (Meigen) (one male, one female). They thus emerged 52 days after the death of the woman in question. Later, during the period 29 September to 6 October, three males and three females of *C. vicina*, as well as three males and 12 females of *L. sericata* emerged. For the period 11 October to 7 November, the rearing cage was placed outdoors where the temperature then dropped below zero. Later, when the cage was again at room temperature additional specimens were released from diapause and during the period 2 January to 11 February 1965 four males and three females of *L. sericata* emerged. The diapause had at least partly occurred in the larval stage, because one larva was seen crawling on 26 January. It pupated on 28 January and emerged as an adult fly on 11 February.

Almost simultaneously with the emergence of the first fly in the rearing experiment, flies also emerged in the autopsy room (*L. sericata* – one male on 24 September, one male on 6 October; *C. vicina* – one male on 28 September, one male on 30 September, one male on 6 October and one male, two females on 8 October). When the refrigerator in the autopsy room was defrosted on 9 October, six males and eight females of *C. vicina* emerged. The larvae from which all these flies had developed had obviously crept into crevices in the autopsy room and refrigerator during the handling of the cadaver.

Conclusions based on the entomological findings. A sample from a dead human body containing full-grown larvae of *L. sericata* and *C. vicina* indicates to the entomologist the following facts.

1. The woman in question had been dead more than 7–8 days, because these flies oviposit on cadavers on about the second day after death and development to mature migrating larvae takes 5–6 days at the prevailing temperature – less in warmer climates (A. A. Green, 1951; Kamal, 1958; Campbell & Black, 1960; Magy & Black, 1962; Rohe *et al.* 1963; Rohe *et al.* 1964).

2. The cadaver had been in southern Finland, because the range of *L. sericata* is mainly restricted to this part of the country (Nuorteva, 1958, 1959*a*, 1961*b*; Nuorteva *et al.* 1964).

3. The cadaver had been in a city, or alternatively on a small archipelago islet, because in Finland *L. sericata* occurs only in cities (Nuorteva, 1959*a*, 1961*b*, 1963; Nuorteva *et al.* 1964) and on small archipelago islets (Nuorteva, 1966*c*).

4. The cadaver had been in sunshine, because *L. sericata* does not usually oviposit on objects with a temperature below 30°C (Cragg, 1956).

Validity of the conclusions. All the entomological conclusions are valid as tested against the known facts. The situation of the cadaver in sunshine was first suggested by the entomological findings and was later confirmed by the policeman who inspected the case. The conclusion that more than 7–8 days had elapsed since death is as such true but valueless, because death had in fact occurred 30 days before the detection of the cadaver. Obviously, the closed windows had delayed the advent of the attractant odour to the sensory organs of the blowflies. If the policeman had looked for the presence of puparia or adult flies in the flat, it might have been possible to draw more correct conclusions. (The development of the flies in the prevailing conditions would have taken about three weeks. Adult flies and puparia were thus probably present in the flat. The finding of newly emerged flies would have reduced the error to one week.)

14. *The badly burnt body (Nuorteva et al. 1967)*

Twenty-five kilograms of the badly burnt remains of a man were found on 25 August 1965 in an old concrete pill-box on the island of Vasikkasaari (south Finland). The soft parts of the remains were swollen, but without noteworthy signs of decay. Autopsy revealed that death had been due to heart failure and this had presumably led to the accidental spread of fire to the petrol cans in the pill-box. The man in question had been seen alive on 16 August 1965 and death had obviously occurred soon after this. The burnt body had thus been in the pill-box for about 8–9 days. When it was found, numerous fly larvae were observed in the ears, eyes and mouth. The length of the larvae varied between six and 16 mm.

The larvae were reared at low room temperature. The first specimens (seven male, three female) of *Calliphora vicina* Robineau-Desvoidy hatched on 14 September 1965, i.e. 29 days after death of the man in question. During the period 27 September to 2 October, 117 specimens (53 male and 64 female) of this species reached the adult stage. In addition, three specimens of *Fannia canicularis* L. (det. L. Tiensuu) hatched during the period 27 September to 2 October.

Conclusions based on the entomological findings. A sample from a dead human body containing larvae of *C. vicina* which reached the adult stage on 14 September indicates to an entomologist the following facts:

1. The death of the man in question occurred before 24 August because the minimum time for the oviposition plus development of *C. vicina* in the conditions of south Finland is 21 days.

2. The death of the man in question had occurred before 18 August, because fly oviposition and development to a full grown larva takes about 7–8 days (A. A. Green, 1951; Kamal, 1958).

3. The body had been in shadow, because the heliophilic *Lucilia* had not oviposited on it (Nuorteva, 1963) although they are in full activity in the middle of August in south Finland (Nuorteva, 1959a, 1959b, 1959c, 1961b, 1964).

Validity of the conclusions. All entomological conclusions were validated by the known facts. The determination of the time of death was in this case more accurate than in Case no. 13. This is due to the fact that the time that elapsed between death and the detection of the dead body was not much longer than the duration of development of *C. vicina*.

15. *The body in the bed (Nuorteva et al. 1967)*

The body of a woman was found in her bed in a flat in the centre of Helsinki, Finland on 1 September 1965. Death had obviously occurred on 10 August, because newspapers had not been removed since that day. This was a case of suicide with sedatives. The cadaver was moderately decayed and greenish. The skin surface was loose and the viscera decayed. Blowfly larvae emerged from the orifices of the body. Autopsy was performed after refrigeration for one day and the fly larvae were collected on this occasion. The larvae were in bad condition at the start of rearing, except for two small specimens. Rearing at room temperature yielded two small imagines of *Calliphora vicina* Robineau-Desvoidy on 27 September.

Conclusions based on the entomological findings. The detection of full grown larvae of *C. vicina* on a dead body on 1 September indicates to an entomologist the following facts:

1. The woman in question had been dead more than 7–8 days, because fly oviposition occurs on about the second day after death and development to mature migrating larvae takes about 5–6 days (A. A. Green, 1951; Kamal, 1958).
2. The cadaver had been in shadow, because the scotophilic *C. vicina* had oviposited on it, but *Lucilia* species had not (Nuorteva, 1963), although they were still active (Nuorteva, 1958, 1959a, 1959b, 1959c, 1961b, 1964).

Validity of the conclusions. Both entomological conclusions were validated by the known facts. The conclusion that more than 7–8 days had elapsed since death is true as such, but of little value, because death had in fact occurred 20 days before the discovery of the cadaver. A better result would obviously have been obtained if living full grown larvae had been taken for rearing, or if puparia or adult flies had been collected by policemen in the flat of the woman in question.

16. *The floating body in the sea (Nuorteva et al. 1974)*

The dead body of a man floating with a life-belt was found in the open Baltic near the Swedish island of Öland on 4 June 1966. He was from a Finnish ship which had gone down in the Baltic on 14 January 1966. The man had thus been dead about four and a half months. The internal organs of the body were badly decayed and the soft parts of the face and chest had decayed away. Elsewhere adipocere formation had occurred. The body was taken to Helsinki, and an immediate autopsy was performed on 10 June 1966 when fly larvae 10–12 mm long were detected in the chest. An attempt to rear the larvae was unsuccessful. Two larvae preserved in ethanol were determined. They represented the species *Coelopa frigida* (Fallén) of the family Coelopidae.

Conclusions based on entomological findings. *Coelopa frigida* is a fly confined to wrack on the sea shore (Backlund, 1945; Remmert, 1965). Its occurrence on the dead body thus indicated that at some stage the body had been by the sea shore. Because the Baltic is still rather cold in May and early June, the larvae may have been as much as 2–3 weeks old. Oviposition had evidently occurred in the first half of May. The absence of blowfly larvae supports this assumption. Their attractions to carcases on the shores of the Baltic does not start until the latter half of May (Nuorteva & Laurikainen, 1964). It may be concluded, therefore, that the dead sailor in his life-belt had floated to the immediate vicinity of a shore during the first half of May.

Validity of the conclusions. There is no evidence as to how the man had floated in the Baltic, but the conclusions are indisputable. In the present case the entomological conclusions had no forensic importance, but they demonstrate the forensic value of *Coelopa frigida* as a sign that a dead body has at some stage been on the sea-shore. They also demonstrate the feasibility of estimating time relations in accordance with the phenology of blowfly oviposition.

17. *The partly submerged woman in the sand-pit (Nuorteva et al. 1974)*

The dead body of a woman was found in an old sand-pit in the city of Helsinki, Finland on 27 June 1964. The sand-pit was filled with water and the body was partly submerged and covered by bits of board. Presumably, it had earlier been completely under water. It was largely decayed, but still retained its original shape. The soft parts of the face had decayed away. Adipocere formation had occurred. At autopsy on 29 June 1964, some fly larvae and puparia were detected in one hand. The larvae were dead and dry by the time they reached the Zoological Museum for examination. In spite of this it was possible to see that three of the larvae belonged to the genus *Muscina* and one to the genus *Fannia*. No flies or fly parasites had emerged from the puparia. No blowfly larvae or water insects were found.

Conclusions based on entomological findings. The occurrence of larvae and unhatched puparia in the body indicate that it had been accessible to flies for at least one week, because in the prevailing cold micro-climatological conditions the development of *Muscina* fly from egg to puparium takes about that time or a bit longer (Thomsen & Hammer, 1936; Nuorteva & Häsänen, 1972; Nuorteva, 1974). The absence of blowfly larvae, which attack corpses in their initial phases of decay, indicated that the body had not been accessible to blowflies at that stage. Meanwhile, decay had advanced so far that the corpse was no longer attractive to blowflies (cf. Bornemissza, 1957), but was attractive to *Fannia* and *Muscina*. Hence it was concluded that the body had been completely submerged for a comparatively long time.

Validity of the conclusions. Police investigations showed that the woman in question had been murdered in the middle of July 1963 and had been hidden by the murderer in the sand-pit. Therefore the conclusion based on the absence of blowflies, although valid, had only a low degree of accuracy. The conclusion based on the occurrence of larvae or puparia of the genera *Fannia* and *Muscina* was indicative of the time during which parts of the dead body had been accessible to fly oviposition (i.e. had been above the water surface), but was of no significance for the determination of the time of death. Obviously, the hand had first emerged from the water and flies had then oviposited on it. The other parts that were above the water surface when the body was detected had no fly infestation.

18. *The case of the blood-stained shirt (Nuorteva, 1974)*

In August 1972 a very complicated crime, including two different murders and one knifing, occurred in the suburbs of the city of Helsinki, Finland. During the disentanglement of this series of crimes, a policeman found a blood-stained shirt in a plastic bag in an outdoor refuse bin. About two hours before the police arrived on the scene, a man had transferred the shirt in the bag to the bin from a nearby house where one of the murders had been committed. Numerous fly larvae and some fresh puparia were found on the shirt, which was wet and smelt of decaying blood. The police wanted to know the time of the murder or knifing. They were also uncertain whether the shirt had been stained with blood during a murder in one of the houses near the bin or during an earlier murder, which had occurred in another locality.

A search was made for full-grown fly larvae and puparia in the house, especially in the kitchen, where the plastic bag with the bloody shirt had been lying since the time of the murder. None was found. It was also noteworthy that there were no fly larvae feeding on the ample amounts of dry blood found in one of the other rooms.

The bloody shirt with the larvae was placed in a rearing can of the type described by Nuorteva, *et al.* (1967) and the culture was kept in a cool room at a temperature of about 16°C. The laboratory environment was obviously very similar to the true case environment. Adult flies emerging in the culture were counted every day. The results are given in Fig. 48. In all, 104 fly specimens belonging to four different species emerged. Specimens of *Muscina stabulans* Fallén emerged in two separate waves, whereas the other three species had only one period of emergence each.

In the present case, evidence existed that the blood-stained shirt had been exposed to flies in the rubbish bin just before it was found by the policeman. The fly eggs and youngest larvae in the shirt obviously resulted from this exposure, but the shirt also contained older larvae. Therefore, oviposition had also occurred in the house where the murder was committed. There was also evidence that the shirt had been wrapped in the plastic bag in such a manner that flies had no access to it for some time before it was transferred to the rubbish bin. It was thus to be expected that flies would emerge in two waves when the larvae were reared in the laboratory.

In fact, the results (Fig. 48) showed two waves of emergence of the species *Muscina stabulans*. The other three species had only one wave of emergence. It was obvious that the later wave of *M. stabulans* specimens dated from the oviposition known to have occurred in the rubbish bin on 24 August. The first specimens of this wave emerged 21 September, and thus the minimum time of development in the prevailing conditions was 28 days. The first specimen of the first wave of emergence appeared on 8 September. From the above-noted minimal development time of 28 days, it is possible to calculate that the shirt had become stained with blood in which flies had laid eggs on August 11th. This coincided with the date of the knifing, as shown later by police investigations.

The right conclusion was reached by making use of the occurrence of the two clearly separate waves of emergence of *M. stabulans*. Such a possibility does not often occur.

19. *The headless body case*

In September 1983 the headless body of a young woman was found hidden in gorse and bracken in Devon. Many full-grown larvae and puparia of *Ophyra* were found in clothing from the body, but only a few larvae and puparia of *Calliphora*.

The absence of significant numbers of blowfly larvae and lack of evidence of their

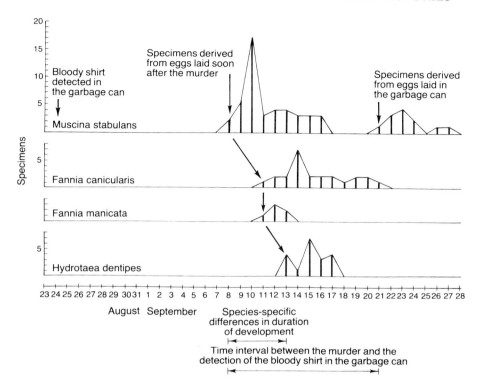

Fig. 48. Case history 18: daily numbers of flies emerging from a bloodied shirt in a rearing can. Two peaks in the emergence curve of *Muscina stabulans* indicated that this species had oviposited on the shirt on two occasions, the other flies only on the first occasion (from Nuorteva, 1974).

feeding in the natural orifices or gunshot wounds on the corpse suggested that the body had been kept elsewhere, probably indoors, for several months and only recently placed on the site where it was found. The good state of preservation of the internal organs (which misled the pathologist to estimate the time of death as 7–10 days), coupled with the presence of *Ophyra*, suggested that the storage place was warm and dry. The presence of the few *Calliphora* larvae and puparia suggested either that the body had been on site for some 20 days or so and being in a dry state had only attracted a few blowflies, or that the head had perhaps been exposed and available to blowflies wherever it was stored and removed on site when a few larvae had crawled off onto the body. When the head was subsequently found it contained several larvae and puparia of *Calliphora*, but only one *Ophyra*, which suggested exposure and subsequent detachment when the differing maggot populations of head and body were then established.

Subsequent confession by the murderer established that the victim had been shot and kept in a sauna room for five months, then dumped at the edge of the wood where the body was found. The head had been removed on site then brought back and kept in a plastic bag in the boot of a car.

Diptera

Flies and maggots

The order Diptera contains the true flies, of which typical adults have only one pair of wings, the second pair being modified as halteres or balancing organs (Fig. 49). The blowflies, houseflies, horseflies, craneflies, mosquitoes and midges belong to this group. About 100 000 species are known to science and many more await discovery and description. In Britain alone, over 6000 species have been recorded.

Flies are probably the most important group of insects affecting man and are certainly the most important carrion insects. The larvae are grubs or maggots, which live in a very wide range of substances and some are aquatic. Some have distinct heads, but no distinct thoracic legs, at most only abdominal prolegs.

The external structures of flies and their larvae are illustrated sufficiently herein to facilitate use of the Keys given, but fuller morphological accounts will be found in the works of Oldroyd (1970b), Colless & McAlpine (1970), McAlpine et al. (1981) and Cole & Schlinger (1969). Very readable accounts of the biology and ecology of flies are provided by Oldroyd (1964) and Stubbs & Chandler (1979). A general account of flies, in German, is given by Hennig (1973). Keys to families of Diptera larvae are provided by Malloch (1917), Hennig (1948–52), Oldroyd & Smith (1973), Brindle & Smith (1979) and Teskey (1981).

References to particular families are cited in the sections that follow and other specialist keys to genera are cited in Kerrich et al. (1978) for Britain and northern Europe, and in Hollis (1980) for the World. The bibliography in the Diptera catalogue for each zoogeographical region (Fig. 46) should also be consulted for literature to species level as follows: Palaearctic (Soós, 1984–), Nearctic (Stone et al., 1965), Neotropical (Papavero, 1966–), Afrotropical (Crosskey, 1980), Oriental (Delfinado & Hardy, 1973–77), (Australasia and Oceania (Steffan & Evenhuis, in prep.).

Identification key to adult flies visiting carrion

[N.B. This is *not* a key to *all* Diptera families and *will only work* for flies specifically feeding or ovipositing on carrion, or whose larvae have been reared from carrion.]

1 Antenna long and slender, composed of two basal segments and a flagellum of at least six, usually more, distinct segments (Nematocera) (Fig. 56) 2
— Antenna short and stouter, composed of two usually shorter basal segments and a third usually larger segment which sometimes has a bristle-like, arista above formed from fusion of the elements of the flagellum which may still be visible (Brachycera) (Figs 57, 58) 4

2 Wings long, narrow and pointed, often held roof-like over back when at rest; moth-like flies clothed in long hairs (Fig. 292) .... **Psychodidae**
— Wings broader and held otherwise; flies not clothed in long hairs 3

3 Larger very delicate brownish flies with long slender legs and quite complex wing venation (Fig. 291) .... **Trichoceridae**
— Smaller blackish flies with simpler venation and a characteristic fork (f) in the distal half of the wing (not carrion feeding but may be attracted to fungal hyphae on carcases in damp situations; Fig. 394) .... **Sciaridae**

4 Wing with a small (discal) cell in the centre (Fig. 393,d); antenna with flagellum (Fig. 57)
 .... **Stratiomyidae**
— Wing without a small cell in the centre and the antenna consisting of only 3 segments + an arista (Fig. 58) 5

5 A brownish hairy honey bee-like fly with a characteristic wing venation (Fig. 295)
 .... ***Eristalis*** **(Syrphidae)**

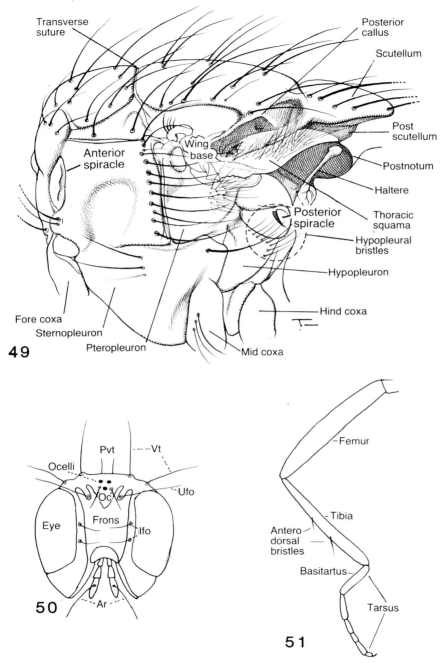

Figs 49–51 Taxonomic characters of adult flies (Diptera). 49, thorax of blowfly (*Calliphora*) in side view. 50, head of fly from front showing bristles; pvt = postverticals, vt = verticals, oc = ocellars, ufo = upper fronto-orbitals, lfo = lower fronto-orbitals, ar = arista. 51, leg of fly.

70   A MANUAL OF FORENSIC ENTOMOLOGY

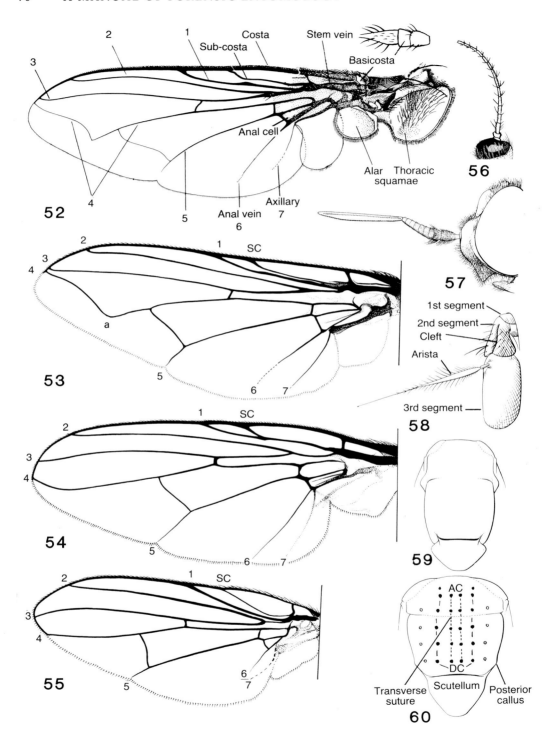

— Fly not honey bee-like 6
[If small and of yellow 'furry' appearance see Fig. 310, Scathophagidae, usually on dung.]
6 Wings with veins running simply and subparallel to wing margin (Figs 293, 294) .... **Phoridae**
— Wing venation more complex with at least one cross vein 7
7 Second antennal segment without a longitudinal cleft on its upper outer edge; thorax usually without a transverse suture and posterior calli undeveloped; squamae usually small (Acalyptratae) (Figs 296–301) 8
— Second antennal segment with a distinct longitudinal cleft on its upper outer edge (Fig. 58); thorax usually with a distinct transverse suture and posterior calli developed (Fig. 60, squamae usually large (Calyptratae) (Fig. 52) 18
8 Costa (first wing vein) complete 9
— Costa with at least one break (Figs 300, 397, cb) 12
9 Small bare black flies with rounded head, abdomen constricted into a 'waist'; often a black spot at wing tip (lacking in *Nemopoda*); usually walking and wagging their wings (Fig. 296) .... **Sepsidae**
— Not as above 10
10 The two bristles (postverticals) on top of the head immediately behind the ocellar triangle converging or crossed 11
— Postvertical bristles parallel (Fig. 50) or divergent .... **Dryomyzidae**
11 Anal vein reaching hind margin of wing as a dark shadow and anal cell closed by a straight crossvein making an acute angle towards the wing margin (Fig. 52); bristly flies on seashore or seaweed .... **Coelopidae**
— Anal vein bluntly shortened and anal cell with the outer border convex and never with an acute angle .... **Lauxaniidae**
12 A break in the costa some distance before point where vein 1 reaches costa 13
— A break in the costa very close to point where vein 2 reaches the costa (other breaks may also be present) (Fig. 397) 15
13 Postvertical bristles diverging, parallel (Fig. 50) or absent 14
— Postvertical bristles convergent or crossed .... **Heleomyzidae**
14 Scutellum long and flattened; antennae vertical, lying in deep grooves in the face (Fig. 301) .... **Thyreophoridae**
  [N.B. This family is now included in Piophilidae by some authors.]
— Scutellum normal (Fig. 60). Antennae inclined forwards, not lying in deep grooves in face (Fig. 299) .... **Piophilidae**
15 One or more of the anterior bristles on top of the head (fronto-orbitals) are directed inwards (Figs 50, 396, fo) .... **Milichiidae**
— No inwardly directed lower fronto-orbitals 16
16 Hind basitarsus (first tarsal segment) shortened and thickened; apical section of vein 4 (and sometimes vein 5) not pigmented; small black or brown flies (Fig. 311) .... **Sphaeroceridae**
— Hind basitarsis not shortened and thickened (Fig. 51) 17
17 Small yellow to pale brownish flies with red eyes; arista 'feathered' with rays above and below and ending in a characteristic fork (Fig. 300) .... **Drosophilidae**
— Darker greyish to dark brown flies; arista, when feathered, with rays confined to the upper surface (often on shore lines, either marine or freshwater) .... **Ephydridae**
18 Hypopleuron (sclerite on side of thorax below posterior thoracic spiracle) bare or with only soft hairs 19
— Hypopleuron nearly always with a row of distinct bristles (Fig. 49) 21

Figs 52–60 Taxonomic characters of adult flies (Diptera). 52, wing of *Calliphora vicina* (bluebottle); 53, wing of *Musca domestica* (housefly); 54, wing of *Muscina stabulans*; 55, wing of *Fannia canicularis* (lesser housefly); 56, antenna of female mosquito; 57, antenna of Stratiomyidae; 58, antenna of blowfly (Calliphoridae); 59, thorax of an acalyptrate fly; 60, thorax of a calyptrate fly showing bristles (ac = acrostichals, dc = dorsocentrals).

19 Lower (thoracic) squama reduced, not projecting; frons (top of head, Fig. 50) without crossed bristles (usually found on dung but may be predaceous on carrion flies; Fig. 310) .... **Scathophagidae**
— Lower squama usually conspicuous (Figs 49, 52) though sometimes less projecting than upper ones; frons often with crossed bristles 20

20 Wings with anal vein (6) extending to wing margin .... **Anthomyiidae**
— Anal vein (6) never reaching wing margin 21

21 Wing vein (7) curved forward in front of apex of vein (6) (Figs 55, 309) .... **Fanniidae**
— Wing vein (7) not curved forward (Figs 53, 54, 307, 308–315) .... **Muscidae**

22 Metallic blue, blue-black or green flies (Figs 302–305, see separate Keys to Genera) .... **Calliphoridae**
— Grey flies with black stripes on thorax and chequered abdomen (Fig. 306) .... **Sarcophagidae**

Identification key to fly maggots on carrion

1 Larvae with an obvious head which is sclerotized and clearly differentiated from the rest of the body 2
— Larvae without an obvious head, which merges with the rest of the body although the mouthparts may be obvious 4

2 Body shagreened, i.e. roughened, like shark skin, and flattened, mouthparts moving vertically, parallel to each other like a pair of hooks (Fig. 73) .... **Stratiomyidae**
— Body not shagreened, mouthparts moving horizontally so that tips can be brought together 3

3 Posterior spiracles surrounded by four fleshy lobes (Fig. 64) .... **Trichoceridae**
— Posterior spiracles situated at the end of a short siphon. Body usually with distinct sclerotized plates (Fig. 70) .... **Psychodidae**

4 Hind spiracles close together and situated at the end of a long respiratory siphon (Fig. 74)
Eristalis (**Syrphidae**)
— Hind spiracles distinctly separated and even if situated on a joint tube then this is forked so that the spiracles are separated 5

5 Larvae with obvious fleshy or spinous processes dorsally and/or laterally 6
— Larvae without such processes though the integument itself may have strong spines 8

6 Flattened larvae with fleshy processes which are branched laterally, at least basally, and may appear feathery. Posterior spiracles on short stalks (Figs 244–246) .... *Fannia* (**Fanniidae**)
— More cylindrical larvae with unbranched lateral and dorsal processes 7

7 Small white, slightly flattened larvae up to 4 mm long with short processes on the dorsal and lateral surfaces; posterior spiracles on brown, sclerotized tubercles, each with a narrow opening (Fig. 85) .... **Phoridae**
— Larger, more nearly cylindrical larvae, with larger longer pointed fleshy processes laterally and dorsally; posterior spiracles in a cleft on posterior face of anal segment and consisting of flattened plates with three slits (Fig. 230, not British) .... *Chrysomya albiceps* (**Calliphoridae**)

8 Posterior spiracles with one or two slits only. These are first or second instar larvae and are generally difficult or impossible to identify except by association with other mature larvae or subsequently reared adults .... (see illustrations for each species)
— Posterior spiracles with 3 slits. Third stage larvae 9

9 Slits of posterior spiracles strongly sinuous (Fig. 252, 260) .... *Musca* (**Muscidae**)
— Slits of posterior spiracles straight or at most arcuate 10

10 Anal segment with short slender fleshy processes in addition to spiracular processes which project backwards (Figs 19, 149, 176) 11
— Anal segment without backwardly projecting processes, at most with a rosette of fleshy protuberances around anal segment 13

11 Posterior spiracles situated on two fleshy processes 12
— Posterior spiracles not situated upon two fleshy processes (Fig. 149) .... **Piophilidae**

12 Anterior spiracles with lobes, along the side of a central column (Figs 120–122) .... **Sepsidae**
— Anterior spiracles with lobes confined to a fan of long filaments at the tip of an often long basal stem (Fig. 170) .... **Drosophilidae**
13 Posterior spiracles on a forked process (Fig. 158) .... *Teichomyza* **(Ephydridae)**
— Posterior spiracles not on a forked process 14
14 Anterior spiracles with a more or less elongated central axis from which arise lateral processes; spiracular openings short (Figs 124–133) .... **Sphaeroceridae**
— Anterior spiracles otherwise 15
15 Borders of spiracular plate with a complete ring of branched hairs (Figs 95, 96; seashore species) .... **Coelopidae**
— Spiracular plate lacking obvious ring of hairs 16
16 Posterior spiracles with short slits arranged at right angles to each other (Fig. 104) .... **Heleomyzidae**
— Posterior spiracles with longer slits arranged at acute angles 17
17 Anal segment with a rosette of fleshy lobes (Fig. 88) .... **Dryomyzidae**
— Anal lobes, if present, forming only short protuberances 18
18 Posterior spiracles sunk in a deep cavity which (at least in preserved specimens) closes over and conceals them (Fig. 190) .... **Sarcophagidae**
— Posterior spiracles not sunk in a deep pit 19
19 Slits of posterior spiracles straight (Fig. 196) .... **Calliphoridae**
— Slits of posterior spiracles distinctly arcuate (*Muscina*, Fig. 284) or bent distally (*Ophyra*, Fig. 270) or, if straight (*Hydrotaea*), lacking a distinct dark peritreme (Fig. 275) .... **Muscidae**

NEMATOCERA

Trichoceridae

(Figs 61–69, 291)

Trichocera spp.

Adults of Trichoceridae (Fig. 291) are known as 'winter gnats' because the common species, *Trichocera regelationis* (L.), *T. saltator* (Harris) and *T. maculipennis* Meigen, etc., fly abundantly during the winter months, although they occur at lower frequencies throughout the year. These are the flies that form dancing swarms during the late afternoon often in cold weather. They are fragile flies resembling small (*c.* 8 mm long) slender Tipulidae (crane-flies or 'daddy-long-legs'), to which they are closely related.

The larvae are saprophagous and feed on decaying material. *T. saltator* occurs in dung and C. Dahl (1973) found larvae of *T. borealis* Lackschewitz on the skeletons of dead voles. More recently trichocerid larvae have been found to constitute an important element in the carrion fauna during the winter months, when the blowfly fauna is absent. Erzinçlioğlu (1980) found adult Trichoceridae attracted to bait of ox heart in December 1979 and observed the first eggs on 15 December, six days after the bait was put out. The first larvae appeared on 20 January and had left the bait by the end of February. A reared specimen proved to be *T. annulata* Meigen. Broadhead (1980) reported larvae of *T. saltator* found on a human corpse during extremely cold weather with frost and snow early in 1978. The fully-grown larvae were found in considerable numbers on the neck region of a partly decomposed body. Larvae were reared outdoors, pupated during October and adults emerged on 31 October 1979. Laurence (1956) reared this species from cow dung during the colder part of the year, but found that by March the gut was evacuated, concentrations of fat body laid down and development was then delayed until the return of the colder weather.

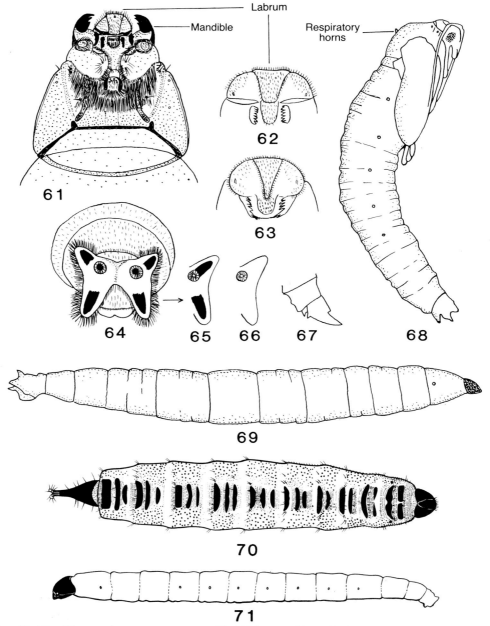

Figs 61–71 Diptera immature stages. 61, *Trichocera hiemalis* (Trichoceridae), larva, head capsule in ventral view; 62, *T. regelationis*, larva, labrum, ventral; 63, *T. maculipennis*, larva, labrum, ventral; 64, *T. hiemalis*, end view of last segment of larva; 65, *T. saltator*, end view of last segment of larva, part; 66, *T. maculipennis*, end view of last segment of larva, part; 67, *T. hiemalis*, posterior end of female pupa, lateral; 68, *T. hiemalis*, male pupa; 69, *T. hiemalis*, larva, lateral; 70, *Psychoda* sp. (Psychodidae), larva, dorsal; 71, *Bradysia* sp. (Sciaridae), larva, lateral.

Keilin & Tate (1940) give a detailed description of the larva and pupa of *T. hiemalis* DeGeer. The presence of a head capsule (Figs 61, 69) and the distinctive appendages of the anal spiracular area (Figs 64, 69) should serve to distinguish the family.

The following key to larvae of *Trichocera* is based on Brindle (1962) and covers the five species most likely to be encountered:

1 The four anal lobes with areas of black pigment (Figs 64–65) 2
- The anal lobes unpigmented (Fig. 66) 3
2 Upper anal lobes pigmented right up to the spiracles (Fig. 65) .... *saltator*
- Upper anal lobes not pigmented as far as the spiracles (Fig. 64) .... *hiemalis*
3 Colour light reddish; posterior spiracles orange-brown .... *annulata*
- Colour yellowish white; posterior spiracles dark brown 4
4 Ventral surface of labrum (Fig. 61) with the area of hairs narrowed behind to a blunt point (Fig. 63)
.... *maculipennis*
- Ventral surface of labrum with the areas of hairs broad behind (Fig. 62) .... *regelationis*

C. Dahl (1973) gives a more technical key to the nine species (of 100 known as adults) so far described in the larval stage. Keys to adults of the British species are provided by R. L. Coe *et al.* (1950) and Laurence (1957). C. Dahl & Alexander (1976) provide keys to the World genera and a check list of species.

Psychodidae

(Figs 70, 292)

The Psychodidae (Fig. 292) are small (*c.* 3 mm) greyish-brown flies clothed with hairs and scales, and hence popularly known as moth-flies. The hairy pointed wings with their characteristic venation of long straight veins serve to distinguish the flies at once. The adults occur throughout spring and summer and are often common in damp places, on tree trunks and sometimes indoors on windows. They run in a jerky fashion, their wings often folded roof-like over their bodies. *Psychoda alternata* Say occurs in vast numbers in the filter beds of sewage works and is commonly called the 'trickling filter fly'. The larvae (Fig. 70) of Psychodidae occcur mainly in liquid or semi-liquid habitats; some breed in cowdung or other excrement and in wet and decaying matter generally.

While the family appears to play no regular part in the faunal succession on corpses, adults have occasionally been noted inspecting dead animal matter. Smith & Grensted (1963) and Beaver (1971) record larvae of *Philosepedon humeralis* (Meigen) from dead snails. My colleague Dr R. P. Lane has reared *P. surcoufi* Tonnoir and other species from a dead sheep. Erzinçlioğlu (1980) records *Panimerus notabilis* Eaton on ox heart. *Clogmia* (= *Telmatoscopus*) *albipunctatus* Williston has been frequently involved in cases of human myiasis (see summary in Smith & Thomas, 1979) and, like *Psychoda alternata*, seems particularly fond of filthy organic conditions.

Payne *et al.* (1968a) record psychodid larvae in buried pigs in the fourth stage of decomposition about three weeks after burial, and adults of *Clogmia albipunctatus* were attracted to floating pig corpses from four to 14 days after death.

Keys to the British *Psychoda* larvae are provided by Satchell (1947) and to adults by Freeman (*in* R. L. Coe *et al.*, 1950).

Sciaridae and Scatopsidae (Figs 71, 72, 392, 394) may occur on carrion but are included under Cannabis Insects (p. 169).

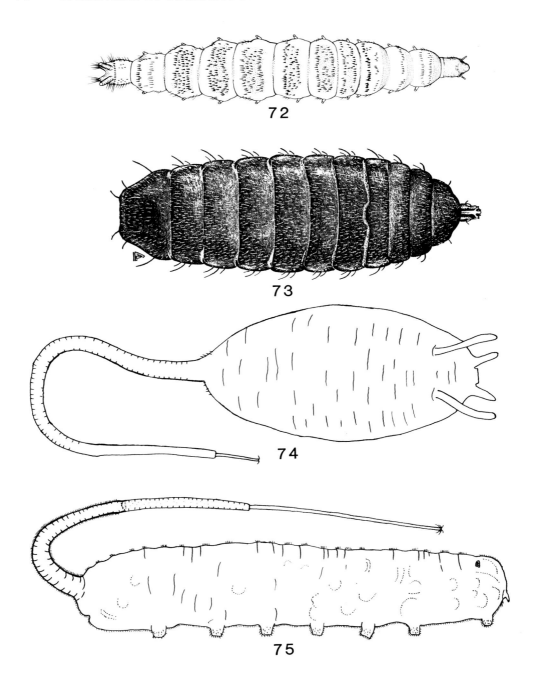

Figs 72–75 Diptera immature stages. 72, Scatopsidae, dorsal; 73, *Hermetia illucens* (Stratiomyidae), dorsal; 74, *Eristalis tenax* (Syrphidae) puparium, dorsal; 75, *E. tenax*, larva, lateral.

BRACHYCERA

Stratiomyidae

(Figs 73, 393)

The Stratiomyidae are small to fairly large (5–15 mm) flies usually found on flowers and leaves. They are sometimes called 'soldier-flies' because of the 'armament' of spines on the scutellum of some species. In colour they range from brilliant metallic greens and purple to black and yellow. They may be distinguished by their generally flat appearance and by the generally small size of the discal cell (d) in the wing (Fig. 393).

The larvae (Fig. 73) are aquatic or terrestrial; the former have an anal tuft of hairs which keeps them afloat when breathing at the surface, and are carnivorous. All stratiomyid larvae may be distinguished by their shagreened appearance caused by a surface layer of calcium carbonate, an adaptation against desiccation when aquatic or semi-aquatic habitats dry out. The immature stages of the terrestrial species are found in rotting wood, under bark, and in dung, lavatories and compost heaps, and will probably feed on any decaying matter of animal or vegetable origin.

Larvae are occasionally found in carrion. Reed (1958) found larvae of *Eulalia, Sargus* and *Ptecticus* under or in dog carcases during the decay or dry stage in wooded areas in the USA (Tennessee). Smith (1975) found larvae of *Microchrysa polita* (L.) from September to November in a fox carcase in England. Payne & King (1972) found *Ptecticus trivittatus* (Say) and *Hermetia illucens* (L.) as scavengers on pig carcases in water in the USA.

Hermetia illucens (Figs 73, 393) is primarily an American species, but has been transported via ships to Europe, Africa, Asia and parts of the Australian region. It is known to breed in human excrement and Dunn (1916) records larvae feeding in great numbers upon the dead body of a man in the jungle of the Panama Canal Zone. No other fly larvae were present on the corpse. Kilpatrick & Schoof (1959) studied the interrelationship of this species with *Musca domestica* in human excrement. The larva is very resistant to insecticides and to concentrated solutions of alcohol and acid, and brine (Bohart & Gressitt, 1951). Larvae have also been found in stored *Cannabis* (see Cannabis Insects, p. 169).

The adults and immature stages of the European Stratiomyidae are monographed by Rozkosný (1982–83).

CYCLORRHAPHA-ASCHIZA

Phoridae

(Figs 76–85, 293, 294)

This is a large family of flies with some 3000 world species, of which 300 occur in the British Isles. They are minute to medium-sized (0.75–8.00 mm), dull black, brown or yellowish flies of humped back appearance, generally bristly and with very characteristic wing venation. They run about in an active erratic manner which has earned them the popular name of scuttle-flies.

Phoridae breed in a very wide variety of decaying organic material, but in addition some are parasites and others develop in fungi. In the larval stage some species are predators. Species of *Spiniphora* appear to specialise in breeding in dead snails, but *S. bergenstammi* (Mik) is also frequently recorded from dirty milk bottles. Several genera are regularly found in vertebrate carrion, e.g. *Anevrina, Conicera, Diplonevra, Dohrniphora, Metopina, Triphleba* and some *Megaselia* species.

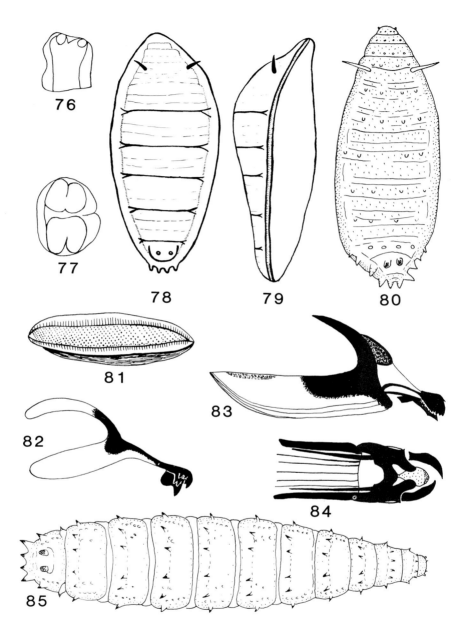

Figs 76–85 Phoridae (Diptera) immature stages. 76, *Megaselia*, anterior spiracle of third instar larva; 77, *Megaselia*, posterior spiracle; 78, *Megaselia rufipes*, puparium, ventral; 79, *M. rufipes*, puparium lateral; 80, *Conicera*, puparium; 81, *Megaselia* egg; 82, *Megaselia*, mouthparts of third instar larva, lateral; 83, *Conicera*, mouthparts of third instar larva, lateral; 84, *Conicera*, mouthparts of third instar larva, ventral; 85, *Megaselia*, whole larva, dorsal. (Figs 76–77 after Kaneko, 1965, *J. Aichi Med. Univ. Ass.* **5**.)

Conicera tibialis Schmitz (Figs 80, 83, 84) is a small (*c*. 1.5 mm) black species known as the 'coffin-fly' because of its association with coffined bodies that have been underground for about a year. The species also occurs on dead moles and in sand martin nests on dead nestlings. In the literature the species has been variously misidentified as *Phora aterrima* (Mégnin, 1887), *Phora vitripennis* (Blair, 1922) and various other species of *Conicera*. The Blair record is of interest as it concerns specimens collected by the well-known pathologist, Dr B. H. Spilsbury, from the hair of a corpse exhumed after ten months' burial at Hay, Herefordshire, in 1921. Colyer (1954a) reviews these earlier records. Mégnin (1887) had concluded that the phorid larvae had moved down through wormholes and other fissures in the soil from eggs laid on the surfaces and this was supported by the usual occurrence with the phorids of the beetle, *Rhizophagus parallelocollis* (q.v.) which normally spends its life history above ground – it was even suggested that the life histories of these insects were in some way interconnected. Others have suggested that the adults burrowed down through the soil or that eggs were laid on the corpses before burial. In many exhumations vast numbers of flies and puparia were found and it appeared that many generations had successfully developed within the coffins. Colyer (1954b, 1954c) reports the occurrence of the species in numbers on the soil surface (in May 1954) above a dead dog buried some three and a half feet deep some 18 months previously. The flies were swarming and copulating in large numbers on the soil surface and digging revealed that adult flies, with wings dried and expanded, were present at all depths from the surface down to the actual corpse. The movement of the phorids appeared to be '"one way traffic", i.e. from the corpse to the surface of the ground'. The soil between corpse and surface was moderately loose and many worm tunnels were present, and young earthworms were developing in the corpse. No eggs were found on the corpse and the May generation had completely disappeared by the middle of June 1954, but adults were again seen swarming over the soil above the dog on 6 August 1954. This activity at the surface occurred in sunshine; in dull weather the flies congregated beneath clods of earth.

In Sweden in March 1952 an exhumation was carried out on a corpse where death had occurred on 19 June 1948. I quote from a translation of part of Ardö (1953):

Larval material found in mid-winter (March). The ground temperature was about 0°C, while the hollow in which the body lay had a higher temperature (4–5°C). The newly hatched phorids were pale grey and the pigmentation appeared to develop very slowly. Thus, 48 hours after emergence, one could easily separate, by its grey colour, an individual from earlier ones, which were all black. The whole of this isolated population emerged in the course of seven days, and the last flies died after a further seven days. One should therefore be able to reckon on an average life of about seven days for a species under normal room temperature.

In spite of a sex-quotient strongly in favour of females, I observed copulation on numerous occasions. On no occasions were these occurrences preceded by any courtship behaviour. It is interesting to see how a fly, which normally lives under the most specialized conditions, as much in regard to temperatures and light intensity as to oxygen supply, can adapt itself to such artificial surroundings as prevail in the bottom of a glass container.

When all the flies were dead, it was possible to observe the newly hatched white larvae, which speedily developed and produced another generation about 50 days after the first adults had appeared from the previous generation. A third generation developed, but was spoiled by being accidently placed in a sunny window where the heat proved altogether too high.

The above experiment can be considered as proof that *Conicera tibialis* is able to produce a number of generations in 'mikrokaverna' without the need for each generation to search out soil in which to swarm and copulate, as in the case of the beetle *Rhizophagous parallelocollis.*

A further quotation from this work:

From an ecological point of view the 'biotope' discussed here is very special. At normal grave depth (more than two metres) the temperature change is only a few degrees, according to maxima and minima measured over about 3 months. The temperature remains in general at about 5°C, and the development of *tibialis* must take place very slowly as shown by the fact that the first flies emerged at a temperature of about 0°C. Further it is important that development can take place independently of season, while the food supply lies at frost free depth.

In Germany, Lundt (1964) found that adult *Conicera* and *Metopina* could reach buried 'flesh' at a soil depth of 50 cm in four days. In the USA Payne *et al.* (1968a), working with pig carcases buried from 50 to 100 cm deep, found *Dohrniphora incisuralis* (Loew) and *Metopina subarcuata* Borgmeier adults had arrived by the third day and fed on fluids from the bloated carcases. Larvae of phorids were actively feeding by the seventh day as deflation of the corpse began.

Disney (1983) keys the British Phoridae. Borgmeier (1968, 1971) catalogues the World species and the appropiate regional catalogue of Diptera should be studied (see introduction to Diptera section (p. 68) in conjunction with Fig. 46).

Syrphidae

(Figs 74, 75, 295)

These are the familiar hover-flies usually seen on flowers or hovering in the sunlight, throughout the world. Many are of bright black and yellow coloration. They are easily distinguished by their characteristic wing venation in which there are outer veins parallel to the wing margin and a false vein (*vena spuria*) between the third and fourth veins (Fig. 295, vs).

Adults may visit carrion seeking moisture, and there is a possibility that the aphid-feeding larvae of some of the common species of *Syrphus* and its allies may accidently drop on to a corpse from overhanging vegetation. These terrestrial larvae are easily distinguished by the fact that the posterior spiracles are fused together and form a short tube.

The only carrion species likely to come to forensic notice is *Eristalis tenax* L. (Fig. 295), which is a common and cosmopolitan fly popularly known as the 'drone-fly'. The adult strongly resembles a honey-bee in its appearance, the droning noise it makes and its frequent occurrence on flowers. The larva (Fig. 75) is popularly called the rat-tailed maggot because of its long telescopically-extensible breathing tube which is an adaptation to life in liquid or semi-liquid media. Any foul water containing rotting organic matter and evil smelling mud provides a suitable medium for this species. These larvae may be found on carrion if there are filthy liquid exudates.

Eristalis tenax is the oxen-born 'bee' (the Greek Bugonia) of the ancients (Osten Sacken, 1894–95) and is illustrated on the tins containing a well-known brand of golden syrup, emerging from a lion carcase with the Samson quotation (Judges xiv, 14) 'out of the strong came forth sweetness'. The larvae have occasionally been recorded in cases of rectal and intestinal myiasis and Zumpt (1965) critically reviews these cases.

Other species of *Eristalis*, or other genera with rat-tailed maggots, may occur on semi-liquid carrion.

Hartley (1961) provides a key to some syrphid larvae.

ACALYPTRATAE

Dryomyzidae

(Figs 86–92)

Dryomyzidae are medium-sized flies (6–10 mm), yellowish or reddish brown in colour and usually found in moist shady areas. They have the wings conspicuously longer than the body. The genus *Helcomyza* is greyish in colour, has a series of costal spines (similar to Heleomyzidae), and occurs on the seashore together with other seaweed frequenting flies such as Ephydridae and Coelopidae.

The egg (Fig. 92) bears two dorsolateral flanges and is particularly well adapted to survival on the type of substrate on which it is laid.

The larvae (Figs 86–91) are found in putrefying matter, especially excrement (including human) and rotting fungi (including the stinkhorn, *Phallus impudicus* Persoon, which smells of carrion – Smith, 1956) and carrion.

Dryomyza anilis Fallén has been recorded from dead shrews (Disney, 1973), a dead fox (Smith, 1975) and a dead pheasant (Smith, 1981). Barnes (1984) describes the biology and immature stages of the species in the USA, where it was successfully reared in the laboratory on dead annelids, molluscs, insects, vertebrates and rotting fungi. Greenberg (1971–73, vol. 1) says that *D. anilis* is 'among the distinctly forest coprophages' with 'a narrow temperature preferendum. It is found on faeces even in dense shrubbery and in shaded and moist places.'

Coelopidae

(Figs 93–99)

The Coelopidae are popularly known as seaweed-flies or kelp-flies as they are normally found on the sea-shore on the kelp or seaweed cast up by the tides.

They are small- to medium-sized (3.0–7.5 mm), flattened, hairy flies with small eyes and deep jowls. They occur in the Palaearctic and Afrotropical regions and in the New World, but are best represented in the Australo-Pacific region. They are not normally regarded as carrion flies, but Nuorteva *et al.* (1974) found *Coelopa frigida* (F.) in a human corpse during the first half of May, before blowflies were active (see Case history 16, p. 65). Thus these flies could have a forensic importance with bodies found on the shore.

The larvae (Figs 93–98) are found in decomposing wrack beds and the puparia (Fig. 99) are either among the wrack or under nearby rocks and/or shingle. While the adults are normally found on the wrack beds they do sometimes make mass migratory movements over short distances and are strongly attracted to aromatic substances, especially trichlorethylene, and have on occasions invaded chemists' and perfumers' shops and hospitals.

Oldroyd (1954) gives an interesting account of the family and Egglishaw (1960) treats the immature stages and biology.

Orygma luctuosum Meigen, a seaweed-fly which used to be included in the Coelopidae, is treated under Sepsidae.

Heleomyzidae

(Figs 100–105)

These are small- to medium-sized (4–10 mm) yellowish, brownish or grey flies generally distinguished by a series of small spines which stand out clearly from the hairs along the

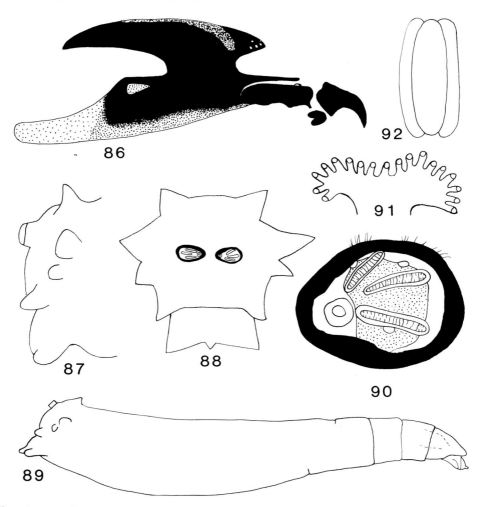

Figs 86–92 *Dryomyza anilis* (Dryomyzidae, Diptera), third instar larva and egg. 86, mouthparts, 87, last segment of larva, lateral; 88, last segment of larva, end view; 89, whole larva; 90, posterior spiracle; 91, anterior spiracle; 92, egg. (Figs 88 and 89 after James, 1947.)

fore-margin (costa) of the wing. On the head are two pollinose stripes on each of which stand one or two (not more) bristles.

The larvae of Heleomyzidae develop in carrion, excrement and various kinds of decomposing organic matter of vegetable or animal origin. They have been reared from rotting potatoes, rotten wood, in rabbit and rodent burrows, in caves where bat excrement is plentiful, in bird and mammal nests and some in fungi. Smith (1956) records five species of *Heleomyza* associated with the stinkhorn fungus (*Phallus impudicus* Persoon) which has a strong carrion odour.

Neoleria inscripta (Meigen) (Figs 100–105) appears to be the heleomyzid most frequently associated with carrion in the British Isles. It has been found on dead cows and rabbits (Parmenter, MS notes), and on a dead fox (Smith, 1975) where its larvae outnumbered all the other Diptera present. On the fox the fly first appeared five days after death and

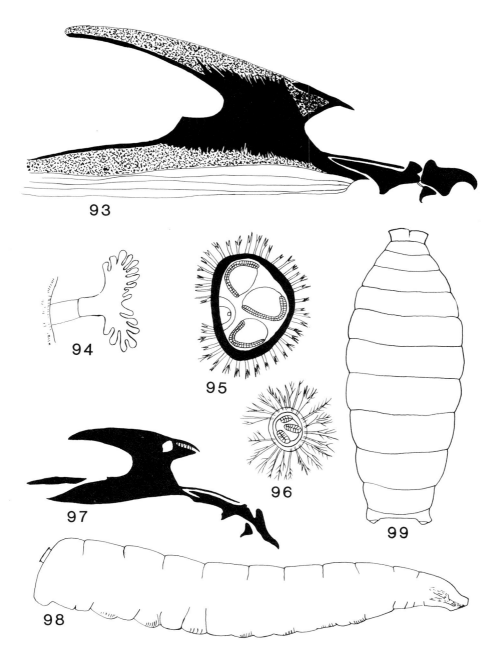

Figs 93–99 *Coelopa frigida* (Diptera, Coelopidae), immature stages. 93, third instar larva, mouthparts; 94, third instar larva, anterior spiracle; 95, third instar larva, posterior spiracle; 96, second instar larva, posterior spiracle; 97, second instar larva, mouthparts; 98, third instar larva, lateral; 99, puparium, dorsal.

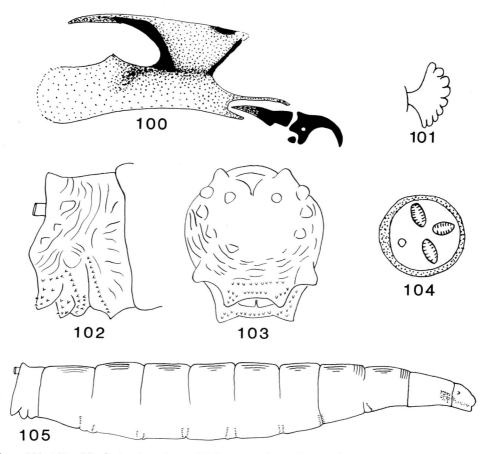

Figs 100–105 *Neoleria inscripta* (Heleomyzidae, Diptera), third instar larva. 100, mouthparts, lateral; 101, anterior spiracle; 102, last segment, lateral; 103, last segment, end view; 104 posterior spiracle; 105, whole larva, lateral.

maggots some three weeks later. *Scoliocentra villosa* (Meigen) and *Oecothea fenestralis* (Fallén) have been recorded in tins of decaying meat sunk in the soil in the entrances of rabbit burrows for the purpose of attracting catopid beetles (Skidmore, 1967). In his studies on buried carrion in Germany, Lundt (1964) found that females of *Morpholeria kerteszi* (Czerny) laid eggs on the surface of the ground and young larvae then move down through the soil to reach the carrion.

Keys to adults of the family are provided by Collin (1943) for British species and by Gill (1962) for North American species and some relevant immature stages are described by Skidmore (1967), Garnett & Foot (1967) and Lobanov (1970).

Sepsidae

(Figs 106–122, 296)

This is another family of small (2.5–3.5 mm) black flies of world-wide distribution, usually seen on excrement among the large yellow dung-flies (Scathophagidae) and the lesser dung-flies (Sphaeroceridae). The habit of wing-waving or wagging as they walk around distinguishes them at once from their companions in this habitat. Members of the common genus *Sepsis* have the further distinction of a black apical spot on the wing (Fig. 296). Adults can sometimes occur in vast swarms on vegetation. The British species may be identified by the keys given by Pont (1979).

Larval Sepsidae feed on a wide range of decaying organic matter, but they are usually coprophagous or occasionally saprophagous, and the genus *Orygma* (which in the past has been included in the family Coelopidae) breeds in the smaller drier beds of decomposing seaweed.

The following key to the larvae of British species is modified from Hennig (1952) and Brindle (1965a) to include *Orygma*. Mangan (1977) keys North American species.

1 Anterior spiracles with distal lobes in a regular fan (Fig. 109); posterior spiracles with U-shaped slits (Fig. 110); usually on the sea shore among seaweed .... *Orygma luctuosum*
- Anterior spiracles more elongated and distal lobes not in a regular fan; posterior spiracles straight or curved, but not U-shaped; usually on dung or carrion 2
2 Anterior spiracles three times as long (or more) as broad, with 12 or more lobes arranged on each side (Figs 120, 122) 3
- Anterior spiracles not greatly elongated, not more than twice as long as broad at base 5
3 Anterior spiracles with 13 lobes, the basal lobes noticeably longer than the distal lobes ....
 .... *Saltella sphondylii*
- Anterior spiracles with 13 or more lobes all about equal in length 4
4 Anterior spiracles with 20 or more lobes (Fig. 120) .... *Themira putris*
- Anterior spiracles with 16 (possibly 13) lobes; posterior spiracles with short elliptical openings (Fig. 122) .... *Nemopoda nitidula*
5 Anal segment not swollen and with prominent ventral lobe (Fig. 115); anterior spiracle with eight lobes .... *Meroplius minutus*
- Anal segment more swollen and without prominent ventral lobes (Figs 112, 113) 6
6 Anterior spiracles with eight lobes .... *Sepsis cynipsea*
- Anterior spiracles with five lobes .... *Sepsis biflexuosa* and *S. thoracica*

On human corpses Sepsidae have been found between the time of caseic fermentation and before ammoniacal fermentation along with Piophilidae, Drosophilidae, etc. On a fox corpse (Smith, 1975) I found that *Nemopoda nitidula* (Fallén) adults first appeared 12 days after death (8 September 1972); second and third instar larvae were found in samples taken 45 days after death (11 October 1972). Erzinçlioğlu (1983) has found them within a few days of death in the north of England. Because of their association with dung, the presence of these flies on or near a corpse should be carefully assessed and the immediate environment carefully studied. Species specifically associated with carrion or human dung are as follows.

Nemopoda nitidula has been reared from excrement, carrion, rotting fungi and logs. It is a very common species, often found on carrion.

Themira putris (L.) has been reared from human and pig dung and birds' nests. It is common around rubbish tips, sewage and farm slurry. Occasionally, enormous outbreaks are recorded in sewage works.

Themira leachi (Meigen) has been reared from human dung and manured soil. Adults are usually found on cow dung.

Themira nigricornis (Meigen) has been reared from human excrement and garden soil.

Sepsis punctum (F.) has been reared from human excrement and also cow and pig dung.

Meroplius minutus (Wiedemann) (= *stercorarius* Robineau-Desvoidy) has been reared from human faeces and accumulations of rabbit dung. Adults are attracted to carrion.

Orygma luctuosum Meigen breeds on the sea-shore, usually on a well-compressed mixture of *Laminaria* and *Fucus*, and may be involved in forensic investigation as has a fellow seaweed-fly (see Coelopidae (p. 81) and Case history 16 (p. 65)).

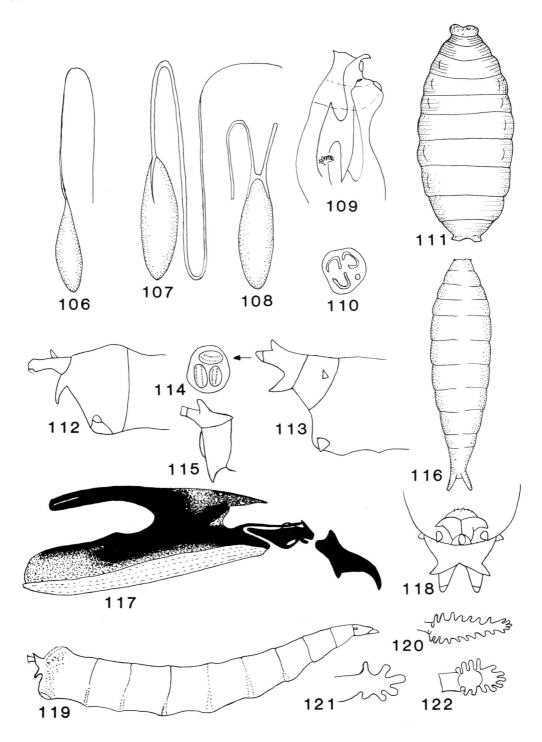

The eggs of Sepsidae (Figs 106–108) have a remarkably long respiratory horn, usually longer than the egg itself.

Sphaeroceridae

(Figs 123–137, 311)

These are small (1.5–3.5 mm), blackish flies of world-wide distribution, often found in company with the more familiar dung-flies on cow pats, etc. (Scathophagidae, Sepsidae). They have a characteristic wing venation (Fig. 311), with veins 5 or both veins 4 and 5 curtailed or faint, and the hind basitarsus (Fig. 51, 311) is usually shortened and dilated. The Sphaeroceridae occur in a variety of situations, in which they also breed, such as dung, carrion, decaying vegetable refuse, seaweed, fungi and in caves. Sphaeroceridae may be expected on human corpses during caseic fermentation if putrid liquids exude (Mégnin's fourth wave), but Erzinçlioğlu (1983) has found them on the day of death and within the first few days of death in the north of England.

Kimosina (Alimosina) empirica (Hutton) (= *Leptocera pectinifera* Villeneuve) has been recorded from a human corpse in Vienna (Duda, 1918: 127, as *L. cadaverina*) and I have seen specimens from a human corpse found in the back of a car in Surrey in 1980. This is reputedly an indoor species, but it has also been found on a dead seal (Lane, 1978) and Pitkin (in press) reared it from rabbit corpses and dog's food (meat). Fredeen & Taylor (1964) also record *K. empirica* in sewage disposal tanks.

In Britain, Parmenter (1952) found *Coproica ferruginata* Stenhammer on a dead mole and Smith (1975) found the following species on a dead fox: *Leptocera fontinalis* (Fallén), *L. caenosa* (Rondani), *Chaetopodella scutellaris* (Haliday), *Coproica pseudolugubris* (Duda), *Spelobia palmata* (Richards), *S. clunipes* (Meigen), *Opalimosina denticulata* (Duda), *Sphaerocera curvipes* Latreille and *Ischiolepta pusilla* (Fallén). R. F. Chapman & Sankey (1955) found *Spelobia luteilabris* (Rondani) on rabbit carcases. Payne *et al.* (1968a) found that the fauna on buried pigs was dominated by ants, Phoridae and Sphaeroceridae Limosininae ('*Leptocera*').

Other species are habitually found indoors and might therefore become involved in domestic forensic cases. *Leptocera caenosa* is commonly found in association with water-closets and can cause a considerable nuisance and health hazard (see Fredeen & Taylor, 1964). *Coproica ferruginata* (may be a nuisance in buildings closely connected with stables (Richards, 1930). Other species frequently encountered on windows indoors, and again worthy of consideration, are *Copromyza similis* (Collin), *Elachisoma aterrima* (Haliday), *Apteromyia claviventris* (Strobl), *Spelobia clunipes* (= *Leptocera crassimana* (Haliday)), *Pullimosina heteroneura* (Haliday) and *Halidayina spinipennis* (Haliday).

It is probable that many of the species associated with animal dung may be attracted to human dung present on a site of forensic importance, or might transfer from animal or human dung to a human or animal cadaver. Some species appear to be more specific in their choice of a particular dung than others (see Pitkin, in press). Species specifically recorded from human excrement are *Leptocera caenosa*, *Sphaerocera curvipes*, *Ischiolepta*

Figs 106–122 Sepsidae (Diptera) immature stages. 106, *Sepsis punctum*, egg; 107, *S. violaceum*, egg; 108, *Orygma luctuosum*, egg; 109, *O. luctuosum*, larval mouthparts; 110, *O. luctuosum*, posterior spiracle of larva; 111, *O. luctuosum*, puparium; 112, *Sepsis punctum*, anal segment of larva; 113, *Nemopoda nitidula*, anal segment of larva; 114, *N. nitidula*, posterior spiracle; 115, *Meroplius minutus*, anal segment; 116, *Sepsis punctum*, puparium in dorsal view; 117, *S. punctum*, larval mouthparts; 118, *Nemopoda nitidula*, anal segment in ventral view; 119, *Sepsis* sp., whole larva; 120, *Themira putris*, anterior spiracle; 121, *Sepsis punctum*, anterior spiracle; 122, *Nemopoda nitidula*, anterior spiracle. All larvae, third instar.

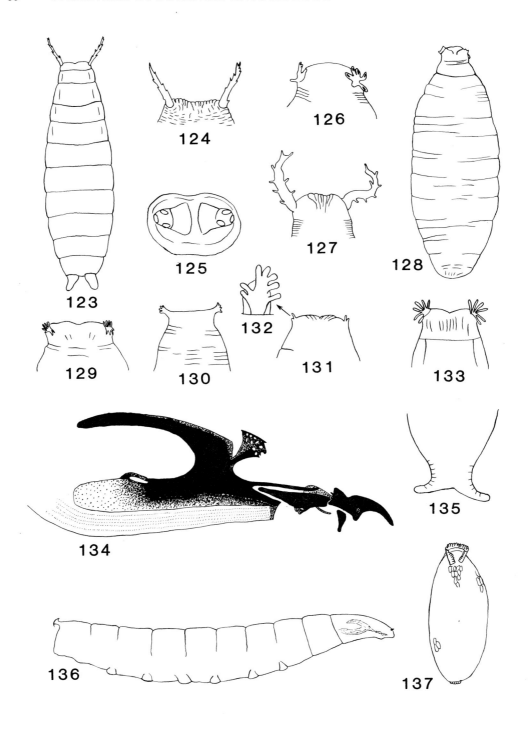

pusilla, Copromyza stercoraria (Meigen), *Coproica vagans* Haliday, *C. acutangula* (Zetterstedt), *Chaetopodella scutellaris* and *Spelobia cambrica* (Richards).

There is no doubt that the wide geographical distribution of many members of this family follows the trail of dung via old caravan routes overland and via the holds of cattle cargo ships.

There is little detailed information on the biology and rates of developments of Sphaeroceridae. Goddard (1938) found that eggs of *Pullimosina heteroneura* hatched after two days, Okely (1974) reports an incubation period of 24 hours for *Spelobia parapusio* (Dahl) and Schumann (1962) also gives 24 hours for *Chaetopodella scutellaris*.

Okely (op. cit.) found that larvae of *S. parapusio* pupated after four days of which 1–2 days were spent in the first instar (at 20°C). Schumann (op. cit.) found that the larval development of *Pullimosina pullula* (Zetterstedt) took 12 days (length of each instar not specified), and that of *C. scutellaris* took five days under laboratory conditions (first instar = one day, second = one day, third = three days).

The duration of the pupal stage appears to be very variable and may be prolonged by unfavourable circumstances, e.g. *Spelobia talparum* (Richards) 5–18 days, *Herniosina bequaerti* Villeneuve 16–18 days (Goddard, 1938) and *P. pullula* 4–12 days (Okely, 1974).

Laurence's breeding experiments (1955) with coprophagous species give some *total* (egg to egg) durations of life histories under natural conditions and similar figures from other sources are indicated as follows:

	days
Limosina silvatica	76–205
Spelobia clunipes	18–100
Chaetopodella scutellaris	20–64
Telomerina pseudoleucoptera	35–60
Opalimosina denticulata	32–45
O. collini	32–50
Halidayina spinipennis	28–34
Pullimosina pullula	22–36 (Okely)
Terrilimosina racovitzai	
Herniosina bequaerti	70–90 (Papp & Plachter, 1976)

Flies probably copulate soon after emergence and oviposition begins 5–12 days later. Parthenogenetic species (e.g. *Ptesemis fenestralis* (Fallén), *Spelobia parapusio* (Dahl)) oviposit 4–5 days after emergence (cf. Goddard, 1938). The length of life of the adults is known only for *Herniosina bequaerti* and *Terrilimosina racovitzai* and is 38 and 62 days respectively.

Hammer (1941) states that *Spelobia clunipes* is not active at night, but Roháček (1982) reports increased activity on warm evenings after rain in *Limosina silvatica, Spelobia clunipes* and *Terrilimosina schmitzi* (Duda), and concludes that temperature and humidity may be more important than light periodicity – but more observations are needed.

Figs 123–137 Sphaeroceridae (Diptera), immature stages. 123, *Spelobia clunipes*, puparium; 124, *S. clunipes*, puparium, detail of anterior spiracles; 125, *S. clunipes*, puparium, showing posterior spiracles; 126, *Ischiolepta pusilla*, anterior end of puparium; 127, *Coproica pseudolugubris*, anterior end of puparium; 128, *I. pusilla*, puparium; 129, *S. subsultans*, anterior end of puparium; 130, *L. caenosa*, anterior end of puparium; 131, *C. stercoraria*, anterior end of puparium; 132, *C. stercoraria*, anterior end of puparium showing single spiracle, enlarged; 133, *T. zosterae*, anterior end of puparium; 134, *Borborus ater*, mouthparts of third instar larva; 135, *T. zosterae*, posterior end of puparium; 136, *S. parapusio*, third instar larva, lateral; 137, *L. caenosa*, egg. (Partly after Okely, 1969; Schumann, 1962.)

Richards (1930) reports that many species are active at night.

The eggs (Fig. 137) are white to pale yellowish, elongate oval and usually dorsally flattened. They are usually laid partly buried in the larval food with only the dorsal or anterodorsal part uncovered. The surface is sculptured and additional structures are often developed, especially around the micropyle, some of which probably serve a similar function to the respiratory horns in some species. The eggs of Limosininae are comparatively large (0.43–0.68 mm).

Few larvae have been described and only one species (*Chaetopodella scutellaris*) is known in all three instars. Larvae are usually white and semi-transparent (Figs 134, 136).

The puparium is usually yellowish to golden brown and is the best known stage (Figs 123–135). The true pupa inside lies in the posterior two-thirds so that the anterior part may frequently be bent.

Most of the work on the immature stages has been focussed on the puparia, but includes descriptions of larval details recovered from them. Papers by Goddard (1938) and Okely (1969, 1974) deal with British species, and those by Deeming & Knutson (1966), Richards (1930), Roháček (1982) and Schumann (1962) are useful for identification of larvae. Pitkin (in press) should be used for the identification of British adults.

The species most likely to be implicated in forensic cases are illustrated (Figs 123–137). *Thoracochaeta zosterae* (Haliday) is included because another 'seaweed-fly' has been implicated forensically, as recorded under Coelopidae.

Carrion studies and others of possible value in forensic investigations and which include Sphaeroceridae are Cornaby (1974), Egglishaw (1961), Mihályi (1967), Nabaglo (1973), Payne (1965), Payne & King (1972) and Teschner (1961). A bibliography of arthropods associated with dung is provided by Kumar & Lloyd (1976). See also Howard (1900) and Steyskal (1957).

Piophilidae

(Figs 138–149, 299, 150–153, 301)

The Piophilidae are small (2.5–4.5 mm), shining black flies of world-wide distribution and are mostly scavengers. The adults are found around carcases, bones, garbage, human excrement, sewage, skins, tanneries, fur stores, etc., and any highly proteinaceous foods of the drier kind.

The larva of *Piophila casei* (L.) is the well-known 'Cheese-skipper', a common stored-product pest on cheese and bacon and, in consequence, of cosmopolitan distribution. The name is derived from the habit of arching the body and grasping its small anal papillae with its mouthparts. When the grip is suddenly released the larva is flung into the air. During such 'skips' the larva can rise two or three inches and cover a distance of several inches, an effective escape mechanism.

Simmons (1927) made a study of the biology of *P. casei* in the USA. Up to 200 or more eggs were laid singly or in groups on the surface of the meat (ham) or packed in crevices. During hot weather the eggs hatched in a day and the larval stages were completed in about five days, then migration commenced. During the migration the skipping habit was much in evidence. Although a useful escape mechanism in times of danger, it was also used during undisturbed migration. However, the most usual and reliable method of locomotion is by creeping. Pupation occurred within three or four hours of termination of feeding and, after 48 hours, 90% of the migrant larvae had pupated. At summer temperatures flies emerged from the pupal stage in 5–8 days. Other workers have found different rates of development due to differing temperatures and feeding media (cheese is inferior to ham as a food for larvae) and Simmons tabulated those up to 1927. Smart (1935) found that larvae could survive temperatures as high as 52°C for one hour and 45°C for an exposure of 24 hours.

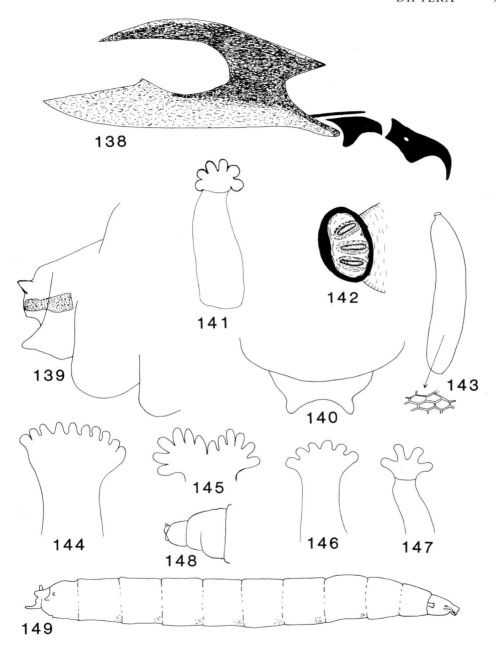

Figs 138–149 *Piophila* (Piophilidae Diptera) larvae. 138, *P. vulgaris*, mouthparts; 139, *P. vulgaris*, rear end in lateral view; 140, rear end in dorsal view; 141, *P. vulgaris*, anterior spiracle; 142, *P. casei*, posterior spiracles; 143, *P. casei*, egg showing surface sculpture enlarged; 144, *P. casei*, anterior spiracle; 145, *P. foveolata*, anterior spiracle; 146, *P. varipes*, anterior spiracle; 147, *P. bipunctatus*, anterior spiracle; 148, *P. bipunctatus*, rear end in lateral view; 149, *P. casei*, whole larva (length 6–8 mm; scale line = 1 mm).

Larvae have been recorded from carrion at a late, drier stage. Colyer & Hammond (1968) found larvae of *P. varipes* Meigen on chicken legs discarded by a fox outside its earth and *P. vulgaris* Fallén on a dead fox on a game-keeper's gibbet.

R. F. Chapman & Sankey (1955) question the 'late arrival' status of *Piophila*. They found adult *P. varipes* on a rabbit corpse three days after death and note that Duffield (1937) found four species of *Piophila* after four days. However, it is important to note that presence of an adult does not necessarily mean that oviposition is occurring and it is the presence of established colonies of larvae rather than the few investigating adults that one associates with a particular 'wave' in the faunal succession. They do emphasize, however, that it is the state of decomposition of a corpse that is all important and, of course, the rate of decomposition is affected by location, temperature and size of a corpse.

The present author (Smith, 1975) found adult *P. varipes* and *P. vulgaris* within four or five days after death on a fox, though only the latter was found in the larval stage some weeks later. Oldroyd (1964) records larvae of *P. foveolata* Meigen (= *nigriceps* Meigen) in the thigh of an exposed human corpse, and Drs M. W. Shaw and W. J. Hendry have sent me specimens of the same species also from the thigh of a human body (found in Scotland on 15 February 1982 and known to have been there since June 1980). In exceptionally fine summer conditions I have seen third instar *Piophila* larvae from corpses known to be only two months old. Nuorteva (1977 : 1078) found *P. foveolata* larvae in a human skull in snow in southern Finland; it was subsequently established that the man had died some nine months previously. *P. foveolata* is usually found in slaughter-houses, meat factories and on poultry farms (Zuska & Lastovka, 1965). See also Forster (1914).

Rondani (1874) first recorded larvae of *P. casei* in an exposed human cadaver in Paris. Mégnin (1894) includes this account in his own discussion of the species and states that *Piophila* is associated when decomposition of the corpse has been underway for 3–6 months and fatty acids and caseic products are present. Johnston & Villeneuve (1897), however, disagree with Mégnin and from their work on exposed cadavers in Canada (one in May and one in August) found that *Piophila* appeared only after the saponification of the fat was well marked. *Piophila* can occur later as illustrated above and, as Leclercq (1969) points out, can exclusively colonize a corpse when the large, more usual, primary cadaver flies such as *Calliphora, Lucilia* and *Sarcophaga* have for some reason been unable to gain access. Mégnin (1894) records an example of a person who had died in his armchair of apoplexy or an aneurism and was found in his room ten months later with myriads of *Piophila* larvae leaping about.

Motter (1898) found remains of *P. casei* in ten of the 150 graves he examined. These graves were from three to ten years old and three to six feet deep. It is not known how the flies entered the graves, but their occurrence suggests that oviposition in the dark is a possibility, and is worthy of further investigation. *Piophila* has also been recorded from Egyptian mummies (Cockburn *et al.*, 1975) and from disinterments (Motter, 1898).

The related families, Neottiophilidae and Thyreophoridae, have now been incorporated into the Piophilidae (McAlpine, 1977). The latter occur on corpses and are included here.

The following key to all British species of *Piophila* so far known in the larval stage is modified from Brindle (1965b) to include *P. foveolata* (= *nigriceps*):

1 Anterior spiracles with four distal lobes and with the basal part of the spiracle narrow (Fig. 147); anal segment narrower, with small lobes on the extreme posterior part (Fig. 148) **bipunctatus**
– Anterior spiracles with more than four lobes and the basal part of the spiracle broader; anal segment broader with a pair of dorsal and ventral lobes (Fig. 139) 2

2 Anterior spiracle with 12 lobes arranged in two groups of six (Fig. 145) .... *foveolata*
– Anterior spiracle with less than 12 lobes arranged otherwise 3

3 Anterior spiracle with ten lobes (Fig. 144); anal segment with ventral lobes much longer than the dorsal lobes (Fig. 149) .... *casei*

- Anterior spiracle with six lobes; anal segment with ventral lobes hardly longer than the dorsal lobes 4
4 Anal segment with the ventral lobes directed posteriorly, and about as long as, but broader than the dorsal lobes (Figs 139, 140) .... *vulgaris*
- Anal segment with the ventral lobes directed ventrally and both about as long and as broad as the dorsal lobes .... *varipes*

Other useful accounts of Piophilidae are given by Hennig (1943, 1948–52).

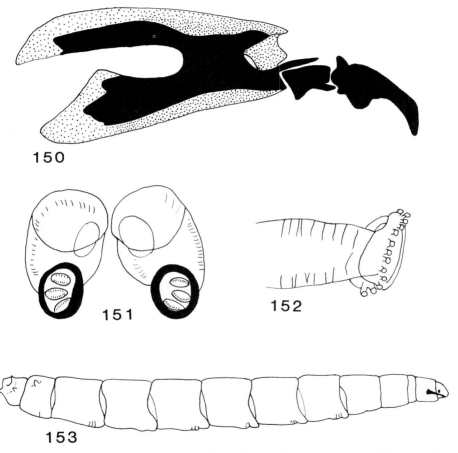

Figs 150–153 *Centrophlebomyia furcata* (Piophilidae, Diptera), larva. 150, mouthparts; 151, posterior spiracles; 152, anterior spiracles; 153, whole larva (length 8–11 mm). (After Freidberg, 1981.)

Those piophilid genera that used to be grouped in the family Thyreophoridae (see McAlpine, 1977) live as larvae in the exposed carcases of large mammals such as horses, donkeys, dogs, deer, etc., which are usually in an advanced state of decay. Only about a dozen species are known and their widely scattered world distribution is suggestive of approaching extinction (Oldroyd, 1964). *Centrophlebomyia furcata*, the only recorded British species, is a yellowish-brown fly, 6–7 mm long (Fig. 301). It has been suggested (Paramonov, 1954; McAlpine, 1977) that thyreophorids are not rare in nature, but rarely collected because they occur at cold times of year when entomologists are not collecting in the field. *Thyreophora cinophila* (Panzer) occurred on corpses immediately after the melting of the snows! Improved conditions of hygiene as applied to the disposal of larger carrion, coupled with the decline in the use of horses and donkeys, may also be responsible for the decline of the group in Europe or their presence should have been detected during the two World Wars. The Palaearctic species had not been recorded since the late 19th Century and were thought to be extinct. However, Freidberg (1981) rediscovered *C. furcata* in Israel on the carcases of goats, cows and sheep in various stages of decomposition during November, December and January and the illustrations of the immature stages herein are based on his work. Eggs are laid on bones and the larvae feed and develop in the marrow, and consequently Freidberg has proposed the name 'bone-skipper' for this species (Figs 150–153). He noted that the adults were particularly attracted to wounds and the skull. What was probably this species has been recorded on anatomical preparations in a medical school in Paris and these flies may therefore be found on human cadavers of forensic interest.

C. anthropophaga Robineau-Desvoidy, another species thought to be extinct, has recently been rediscovered by Michelsen (1983) on the carcase of a horse in Kashmir. Ozerov (1984) describes the larva of *Protothyreophora grunini* Ogerov from 'large mammal' corpses in Russia.

Ephydridae
(Figs 154–159)

A large family (*c.* 1000 species) of small to very small flies, usually of dull brownish or greyish appearance with a very wide mouth opening and strongly arched face. They are usually found near water on shore lines and in marshy places, but occupy an amazing range of very hostile habitats from asphalt lakes, hot springs and salt marshes, to the more favourable sewage and carrion.

Teichomyza fusca Macquart (Figs 155–158) is an almost cosmopolitan, small (5 mm) brownish-olive fly with two grey stripes at the front of the thorax and smoky wings. It is normally found below high-water mark on the sea-shore beneath chalk cliffs, but it also inhabits sewers and the larvae can occur in such vast numbers in liquid excrement that they block sewage and drain pipes. The larvae also occur in woodwork and other materials soaked in urine (animal or human). Oldroyd (1964) records a mass taken from woodwork from a medieval excavation and I have seen numbers of puparia from medieval cess pits (e.g. Greig, 1982). The species appears less common than it used to be, probably because modern toilet systems lack the lime and brine that were so much in evidence in latrines of the past and provided the resemblance to its natural habitat. The only recent specimens I have seen were from pig pens in Sweden. See also Vogler (1900).

Hecamede species (Fig. 159) are tiny (1.5 mm) pale-grey flies with whitish wings which frequent the sea-shore where they run fast over the sand, fly reluctantly and breed in marine rejectamenta such as dead fish. In Guam, Bohart & Gressitt (1951) found *Hecamede persimilis* Hendel abundant on sea beaches and 'on one occasion, the maggots were observed in large numbers in damp foul-smelling sand beneath a human carcase in company with puparia of *Chrysomya "nigripes"* (Calliphoridae) and *Pseudeuxesta prima* Osten-Sacken (Otitidae). Puparia were found with the larvae'. There is one British species, *Hecamede albicans* (Meigen), distributed throughout the western seaboard of the Palaearctic Region and it has been found on dead fish.

Discomyza species (Fig. 154) are small- to medium-sized (2.5–4.0 mm) blackish flies with a broad flat abdomen. They are recorded as breeding in dead molluscs and other carrion. On Guam, Bohart & Gressitt (1951) collected *D. maculipennis* (Wiedemann) in carrion baited traps and from large carcases in the field. They also found them in

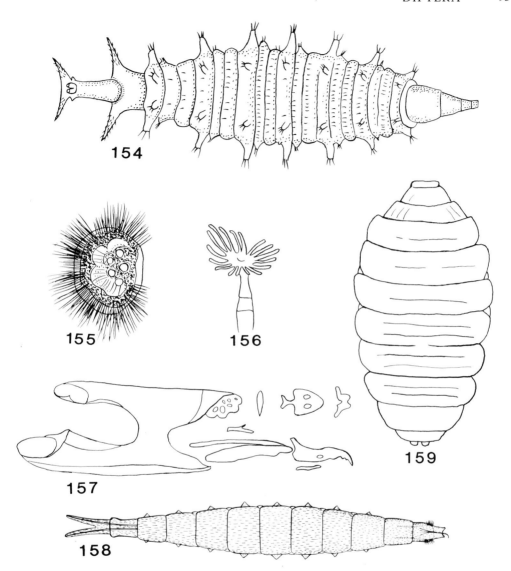

Figs 154–159 Ephydridae (Diptera) immature stages. 154, *Discomyza incurva*, third instar larva (after Hennig, 1952); 155, *Teichomyza fusca*, third instar larva, posterior spiracle; 156, ditto, anterior spiracle; 157, ditto, mouthparts; 158, ditto, third instar larva; 159, *Hecamede persimilis*, puparium (after Bohart & Gressitt, 1951).

company with Phoridae and *Fannia pusio* (Wiedemann) (Fanniidae) in collections of rotting molluscs. Captive flies mated readily and eggs were laid within a few hours and scattered freely, whether food was present or not. The larvae fed readily on liver in the laboratory and were very active; when disturbed, they raised their hind portions in the air. On pupation they remained firmly attached to the food substance. These authors found maggots on a human carcase several weeks old, mostly in deep muscle tissue.

There is one British species, *D. incurva* (Fallén), which has been reared from dead snails and occurs widely throughout the Palaearctic region.

It is likely that, because of their very diverse habits other genera of Ephydridae will become implicated in the forensic investigation of human cadavers, especially on swampy ground or near water, or where particularly harsh environments exclude the normal blowflies.

For further descriptions of the immature stages of Ephydridae see Hennig (1948–1952) and R. Dahl (1969). There is no key to world genera for adults, but Sturtevant & Wheeler's (1954) Nearctic key is very useful. Works on other regions are included in Kerrich *et al.* (1978) for British and northern Europe, and Hollis (1980) for world genera.

Drosophilidae

(Figs 160–176, 300)

Drosophilidae, the lesser fruit-flies or vinegar-flies, are usually small and yellowish-brown (2–4 mm), with a characteristic fork to the tip of the antennal arista, formed from the stem and the last hair (Fig. 300, a). They have a slow, almost hovering flight and are attracted to practically any fermenting substance, in which they breed. They are common in breweries, public houses, pickling factories, fruit and vegetable canneries, canteens and restaurants and frequently occur in sour milk residues in unwashed milk bottles. Many of the 2000 known species have become widely distributed by commercial traffic.

The eggs have long respiratory horns (Figs 171–175). The larvae (Fig. 176) are whitish in colour with prominent posterior spiracles and frequently with prominent anterior spiracles (Fig. 170). The puparium is pale brown with equally prominent spiracles, the anterior one frequently having fan-like tips (Figs 160–164). The immature stages of several cosmopolitan species closely associated with man and that could possibly be implicated in forensic cases are illustrated.

Some species are found on carrion, but it is certainly not the most attractive of baits for this family. However, they may be found when putrid liquids exude.

Drosophila phalerata Meigen, *D. subobscura* Collin, *D. confusa* Sturtevant and *Scaptomyza graminum* (Fallén) have been recorded from the carrion-smelling stinkhorn fungus (*Phallus impudicus* Persoon; Smith, 1956; Nielsen, 1963). Smith (1975) found *D. melanogaster* Meigen on a dead fox. In the USA, Payne & King (1972) found *Drosophila quinaria* Loew and *D. affinis* Sturtevant on the floating remains of pigs immersed in water, and Cornaby (1974) found five species on dead lizards and toads in Costa Rica. *D. funebris* (F.) has been reared from cesspools in France and *D. ananassae* Doleschall has been reared from human excrement. Bohart & Gressitt (1951) record that labels on specimens of *D. busckii* Coquillett in the US National Museum include 'living on cadavers in medical research laboratory'. Nuorteva (1974) found hundreds of larvae of *D. funebris* (in company with *Fannia manicata* Meigen and *Hydrotaea dentipes* F.) in dirty clothes in washbowls during a forensic investigation (Case history 18, p. 66).

Figs 160–176 Drosophilidae (Diptera) immature stages. 160, *Drosophila melanogaster*, puparium, dorsal; 161, *D. busckii*, puparium, dorsal; 162, *D. funebris*, puparium, dorsal; 163, *D. subobscura*, puparium, dorsal; 164, *D. phalerata*, puparium, dorsal; 165, *D. melanogaster*, mouthparts of third instar larva; 166, *D. melanogaster*, enlarged detail of mandible, first instar; 167, *D. melanogaster*, enlarged detail of mandible, second instar; 168, *D. melanogaster*, enlarged detail of mandible, third instar; 169, *D. busckii*, mandible of third instar larva; 170, *D. melanogaster*, anterior spiracle of third instar larva; 171, *D. melanogaster*, egg; 172, *D. busckii*, egg; 173, *D. funebris*, egg; 174, *D. subobscura*, egg; 175, *D. phalerata*, egg; 176, *Drosophila*, third instar larva in ventral view. (Mostly after Okada, 1968 and Shorrocks, 1972.)

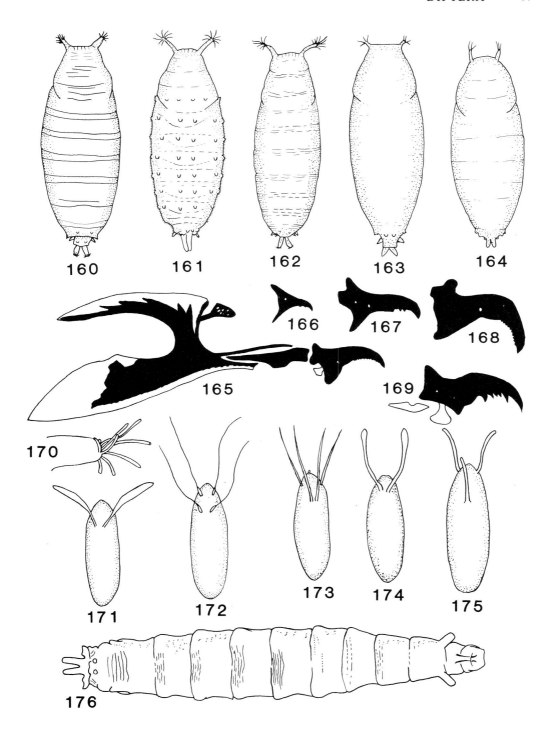

Developmental rates for *Drosophila* are rather rapid; hence their popularity with geneticists. A few figures for *D. melanogaster* demonstrate this.

A female will lay 400–900 eggs at a rate of 15–25 per day, but these are approximate figures only and depend on temperature and other factors. Larvae will develop through to adults as follows:

at	approximate time (in days)
15°C (59°F)	30
20°C (68°F)	14
25°C (77°F)	10
30°C (86°F)	7½

An adult may live from 13 days at 30°C (86°F) to 120 days at 10°C (50°F).

Keys to adult *Drosophila* occurring in Britain (including several cosmopolitan species) are provided by Assis Fonseca (1965). The immature stages of the family are described by Okada (1968). Shorrocks' (1972) and Demerec (1950) should also be consulted (see also Cannabis Insects, p. 169).

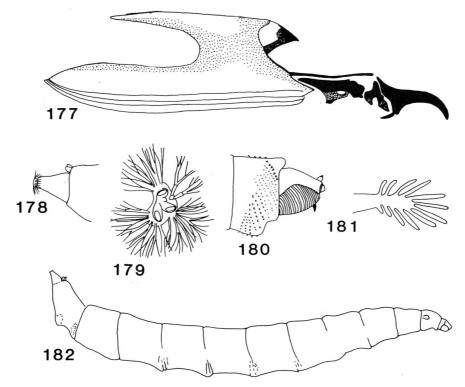

Figs 177–182 *Leptometopa latipes* (Milichiidae, Diptera), third instar larva. 177, mouthparts, lateral; 178, posterior end, lateral; 179, posterior spiracle, end view; 180, anterior end, lateral; 181, anterior spiracle, lateral; 182, whole larva, lateral. (After Hennig, 1956).

Milichiidae

(Figs 177–182, 297, 395–397)

These are small to very small greyish-black or shining black flies with a wide variety of habits.

Madiza glabra Fallén (Fig. 297) is a tiny (2.5 mm) black fly of Holarctic distribution which often occurs in country areas where sanitary arrangements are still primitive. Sometimes adults appear in very large numbers indoors, a phenomenon which has not been fully explained. Oldroyd (1964) suggests that these swarms may result from a mass emergence from some undetected pocket of animal matter.

The life history and immature stages are unknown, but adults have been reported on bodies during 'caseic' or protein fermentation, especially if putrid liquids exude.

Leptometopa and *Desmometopa* are known to breed in human excrement. In Guam, Bohart & Gressitt (1951) reported that Milichiidae were rarely seen on carrion, but some were attracted to human excrement. Howard (1900) records the Holarctic and Afrotropical *Leptometopa latipes* (Meigen) from human excrement. Hennig (1956) described the larvae of *Leptometopa coquilletti* Hendel and what is probably *L. latipes*, and illustrations based on his figures are included here (Figs 177–182).

Desmometopa varipalpis Malloch (Figs 395–397) has a world-wide distribution and has been recorded in association with stored *Cannabis* in the USA (Arnaud, 1974; see also Cannabis Insects p. 169). The larva of *D. tarsalis* has been described by Zimin (1948) and is very similar to Hennig's (1956) illustrations of *L. latipes*. Adult *Desmometopa* attach themselves to large predatory insects such robber flies (Asilidae), reduviid bugs and spiders, and are thus transported until prey is captured, when they join the predator in feeding from the wounds of the prey. Sabrosky (1983) keys World *Desmometopa*.

CALYPTRATAE

Sarcophagidae

(Figs 183–191, 306)

The Sarcophagidae is a large family of cosmopolitan distribution, the members of which are commonly called flesh-flies. The adults are generally large (4–16 mm), silvery-grey and black with a striped thorax, a tessellate or spotted abdomen, and usually strongly bristled and with reddish eyes. The genus *Sarcophaga* is associated with carrion, but this and other genera (e.g. *Wolfahrtia*) are also known to cause wound and intestinal myiasis in man and his domestic animals. These flies, like *Lucilia*, are mostly attracted to carrion exposed in sunlight (A. A. Green, 1951; Smith, 1956) though Payne (1965) found them attracted to pig carcases within five minutes of removal from the freezer and Suenaga (1963) states that *Sarcophaga* will fly in the rain, and in such conditions may be the first arrival at a corpse. *Sarcophaga* usually arrives after the main blowfly sequence, when the carrion is older (D. G. Hall, 1948; Rodriguez & Bass, 1983). The females are viviparous and deposit active first instar larvae. The larvae are characteristic in having the posterior spiracles situated in a deep cavity (Fig. 190).

Sarcophaga species are very similar in appearance and difficult to identify, both as larvae and adults (especially females) and where possible, especially with tropical species, larvae should always be reared through to adults and males sent for specialist examination. Hundreds of species develop in excrement, carrion or any kind of decomposing matter and several species may come before the forensic scientist. The following are species on which some developmental data are available.

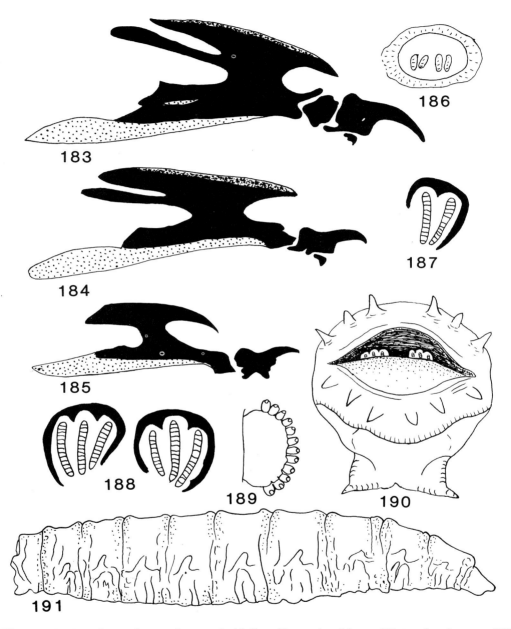

Figs 183–191 *Sarcophaga haemorrhoidalis* (Sarcophagidae, Diptera), larva. 183, mouthparts, lateral, third instar; 184, mouthparts, lateral, second instar; 185, mouthparts, lateral, first instar; 186, posterior spiracle, first instar; 187, posterior spiracle, second instar; 188, posterior spiracle, third instar; 189, anterior spiracle, third instar; 190, posterior view of third instar larva showing sunken spiracles; 191, third instar larva in lateral view.

Sarcophaga haemorrhoidalis (Fallén) (Figs 183–191) is a follower of man and consequently has an almost world-wide distribution. It is commonly called the red-tailed flesh-fly because the male has reddish genitalia. However, other species also possess this character so care must be exercised in identification. This species is mainly a faeces breeder, including human faeces, and in warmer areas, especially where primitive sanitary conditions prevail, adults will commonly occur indoors and are especially attracted to fresh stools. In South Africa (Johannesburg), Zumpt (1965) found that in human faeces the first larval stage lasted about ten hours, the second stage on average 15 hours, the third stage pupated in three days and the adult emerged four days after that. Thus, in favourable summer conditions, adults may appear in eight days from larviposition. Zumpt (1952) found this the commonest *Sarcophaga* species visiting meat bait and human faeces in Johannesburg, together with smaller numbers of *S. tibialis* Macquart, *S. exuberans* Pandellé, *S. inaequalis* Austen, *S. munroi* Villeneuve and *S. inzi* Curran. In the USA E. F. Knipling (1936) found that development from first instar larva to adult took 14–16 days and noted that six females each produced 30–40 larvae.

Sarcophaga hirtipes Wiedemann is widely distributed over the Palaearctic and Afrotropical Regions and breeds in all kinds of decomposing matter, including carcases and human stools.

Sarcophaga misera Walker is a carrion feeder distributed throughout the Oriental and Australian Regions and occurs on many Pacific islands. There has been some confusion over the identity of this species and Lopes (1959) should be consulted for the synonymy in the literature. Alwar & Seshiah (1958) recorded (as *S. dux*) cases of wound myiasis in a circus camel, bullocks and a cow in Madras (India), and gave some information on rates of development. The maggots were deposited some 29–35 at a time and reared on stale meat. The larval period lasted 4–8 days and the pupal period 7–11 days, at 74–94°F. The shortest interval between emergence of the female and larviposition was seven days.

Sarcophaga crassipalpis Macquart is widely distributed over the Holarctic Region and also occurs in South America and Australia. This species usually develops in carrion and C. N. Smith (1933; as *S. securifera* Villeneuve) reared larvae on fresh beef at 80°F in the USA. The first stage lasted 13–16 hours, the second stage 10–30 hours and pupation occurred five or six days from deposition. The flies emerged 10–12 days after pupation and females would commence producing larvae in 8–9 days in batches of 15, normally in the eggshell. One female produces over 100 larvae during her life, which may last up to one month.

Sarcophaga argyrostoma Robineau-Desvoidy is an Holarctic species which has also reached South America, India, the Marshall Islands and Hawaii. Hafez (1940) studied its life history in Egypt and found that at 25°C on meat the first larval stage lasted 30 hours, the second about two days, the third about five days and the pupal stage about eight days. The rate of development varied at different temperatures and in different media. The species occurs in faeces and in cases of myiasis. In Belgium, Leclercq (1976) reports this species from a cadaver in an advanced state of putrefaction. In Guam, Bohart & Gressitt (1951) found *S. gressitti* Hall & Bohart and *S. dux* Thomson under a corpse on a sandy beach. Developmental rates of some other *Sarcophaga* species are given by Kamal (1958; see also Table 5, p. 46).

Sarcophaga carnaria L. (Fig. 306) is a very common species in the Palaearctic region and what was reputed to be this species has the honour of involvement in the first reported forensic case (Yovanovitch, 1888 – see Case histories, p. 56). Draber-Mońko (1973*b*) describes the larva, yet the life history is little known; some claim it feeds exclusively on earthworms, but Greenberg (1971–73, vol. 1) refutes this. Putman (1978*b*) found it on carrion in grassland habitats. This and many other common world species (and related genera such as *Wohlfahrtia*, as yet indistinguishable as larvae) clearly require investigation as to their preferences for carrion, excrement and other decaying matter, in order to establish their precise forensic potential. To assist in the taxonomy in such studies the appropriate Diptera Catalogue (see introduction to Diptera Section, p. 68) should be consulted. The following works (and their included references) will be found invaluable. *Adults*: Emden (1954), UK; Rohdendorf (1930), Draber-Mońko (1973a Palaearctic; Aldrich (1916), Roback (1954), Nearctic; Lopes (1969) Neotropical; Senior-White *et al.* (1940), Oriental; Zumpt (1972), Afrotropical; Lopes (1959), Australia. *Larvae*: C. T. Green (1925), E. F. Knipling (1936), Sanjean (1957), USA; Ishijima (1967), Japan; Mengyu (1982), China; Nandi (1980), India; Cantrell (1981), Australia.

Calliphoridae
(Figs 33–44, 49, 52, 192–240, 302–5)

These are the blowflies and this family includes the genera of greatest importance to the forensic entomologist in estimating the time of death of human cadavers, e.g. *Calliphora* (bluebottles), *Lucilia* (greenbottles). The family contains over 1000 described species and is well represented in all the zoogeographical regions.

A great number of species develop in decomposing organic matter, including carrion, while some are obligate parasites of vertebrates and many develop in other invertebrates, especially arthropods. Norris (1965a) reviews the bionomics of blowflies; Zumpt (1965) treats the species involved in myiasis in the Old World and includes detailed descriptions of the immature stages of many species which also occur in carrion; James (1947) gives a similar but shorter account for the New World; and Smith (1973b) also provides illustrated keys and bionomic information for the important world species.

The taxonomy of these flies is not easy and certainly many more species have potential forensic importance than are recorded in the literature. The forensic entomologist should study in detail the references contained in the appropriate Diptera Catalogue for the zoogeographic region concerned and also those in Hollis (1980). The following more comprehensive works are invaluable introductions containing descriptions, keys and illustrations.

Palaearctic Region: Fan (1965), Kano & Shinonaga (1968), Lehrer (1972), Emden (1954; nomenclature dated, see Zumpt (1956a)), Smith *et al.* (1973).
Nearctic/Neotropical Regions: D. G. Hall (1948), James (1947, 1955), Greenberg & Szyska (1984), Dear (1985).
Afrotropical Region: Zumpt (1956b, 1958).
Oriental Region: Senior-White *et al.* (1940), Kurahashi (1971).
Australasian & Pacific Regions: Kurahashi (1971), Norris (in press), Dear (1986).

More detailed works are cited under each genus.
The following keys should facilitate the identification of the most important Calliphoridae occurring on carrion.

Key to the most important adult Calliphoridae associated with carrion

1 Base of stem-vein (radius) with a row of bristle-like hairs above 2
 – Base of stem-vein bare above (Fig. 52) 5
2 Green to violet-green species, with three prominent black longitudinal stripes on the thorax
 .... (New World) ***Cochliomyia***
 – Green to bluish-black species with at most two narrow longitudinal thoracic stripes 3
3 Thoracic (lower) squamae with fine hairs above (Figs 52, 303) .... (not British) ***Chrysomya***
 – Thoracic squamae bare 4

4 Anterior (prothoracic) spiracle (Fig. 49) with bright orange hair .... *Phormia regina*
  – Anterior spiracle dark-haired .... *Protophormia terraenovae*
5 Greyish flies with three broad, black, longitudinal stripes on thorax and chequered abdomen (Fig. 306; separate family Sarcophagidae) .... *Sarcophaga*
  – Metallic blue or green flies 6
6 Flies of wholly metallic green or coppery-green coloration (Fig. 304) .... *Lucilia*
  – Flies of metallic blue to blue-black coloration 7
7 Abdomen blue with conspicuous overlay of greyish dust (Fig. 302) .... *Calliphora*
  – Abdomen shining blue 8
8 Head black above, face and anterior part of cheeks yellow ..... (Nearctic) *Cynomyopsis cadaverina*
  – Anterior half of head above, whole of face and cheeks, bright orange
    .... (Holarctic) *Cynomya mortuorum*

Key to genera of third stage larvae of the most important Calliphoridae (blowflies) on cadavers

1 Peritreme (Fig. 38, *p*) of posterior spiracle incomplete and not enclosing button (Fig. 38, *b*), which is sometimes indistinct 2
  – Peritreme of posterior spiracle complete, although sometimes weaker in region of button 5
2 Posterior margin of segment 11 *without* dorsal spines; posterior spiracle without a distinct button (Fig. 239) .... (New World) *Cochliomyia*
  – Posterior margin of segment 11 *with* dorsal spines; button distinct or indistinct 3
3 Button indistinct (Fig. 227) .... (Old World, tropics and subtropical) *Chrysomya*
  – Button distinct 4
4 Dorsal spines *present* on posterior margin of segment 10; larger tubercles on upper margin of posterior cavity distinctly longer than half the width of one posterior spiracle (Fig. 222, *t*)
    .... (Holarctic) *Protophormia terraenovae*
  – Dorsal spines *absent* on posterior margin of segment 10; length of larger tubercles on upper margin of posterior cavity less than half the width of one posterior spiracle
    .... (Holarctic) *Phormia regina*
5 Accessory oral sclerite present (Figs 39, 40, 192, 197, 201, aos) 6
  – Accessory oral sclerite absent (except *L. ampullacea*) .... (World) *Lucilia*
6 Mouth-hook with the tooth-like apical portion longer than greatest depth of the basal portion (Fig. 192) .... (World) *Calliphora*
  – Mouth-hook with apical portion as long as greatest depth of the basal portion (Fig. 201)
    .... (Nearctic) *Cynomyopsis*, (Holarctic) *Cynomya*

Genus *Calliphora* Robineau-Desvoidy

These are the familiar bluebottles and are large (6–14 mm), blue, bristly flies (Fig. 302). The genus is best represented in the Holarctic and Australian Regions but two species occur in the Afrotropical Region (*C. croceipalpis* Jaennicke and *C. vicina* Robineau-Desvoidy).

Bluebottles are typically flies of shady situations and were originally probably an essential ecological component of wooded country. Forensically, they are the most important flies involved in cadaver succession in temperate regions. They do not normally fly in the dark, which means that any eggs found on a corpse during the night or early morning were probably laid the previous day, or perhaps earlier in cool conditions. However, Green (1951) notes that in slaughter houses *Calliphora* will fly and oviposit at night though *Lucilia* apparently does not. Extremes of temperature also affect flight activity, but even during the winter individual specimens of *Calliphora* will appear on sunshiny days. While there is little evidence that such individuals will copulate or oviposit, Whiting (1914) claims to have bred *Calliphora* throughout the year in Massachusetts, USA. *C.vicina* can complete its development at 15°C but fly activity ceases at temperatures much below 12°C (Nuorteva 1977). In Britain no new generation is produced between autumn and spring and many of the autumnal adults do not even copulate. However, they may hibernate in suitable sheltered situations, such as outbuildings and emerge during warm sunny spells. The progeny of the final generation overwinters mainly in the prepupal stage, pupates in March and April, and emerges as soon as the minimum ground temperature exceeds 5°C (Wardle, 1927).

Adult *Calliphora* will commence oviposition 4–5 days after emergence, at 24°C (75°F). Payne (1965) records calliphorid oviposition on partly frozen pig carcases, but it is doubtful if oviposition normally occurs at temperatures below 12°C. Nielsen & Nielsen (1946) found that at temperatures below 4°C *Calliphora* eggs did not hatch but at 6–7°C hatching of eggs and larval development occurred. Oviposition on carrion can occur within two days, often within a few hours and, in the right conditions (season and situation), almost immediately after the death of the host. When suitable food is unavailable eggs may be retained in the female's body and living first stage larvae are eventually deposited. Blowfly females have excellent olfactory senses and are very skilful and persistent in seeking suitable oviposition sites. A female will lay up to 300 eggs in smaller groups or a single batch in the natural orifices and crevices in the skin. The eggs are white, banana-shaped and about 1.7 mm long (Fig. 199). When the eggs hatch the young larvae immediately attempt to penetrate the tissues via natural or unnatural openings (i.e. wounds). The developmental rates of the immature stages are shown in Table 5. On reaching maturity the larvae leave the corpse in large numbers and search for suitable pupation sites, usually in the top 2–3 inches of the soil up to a distance of 20 feet from the corpse.

Calliphora species usually arrive first (or with *Lucilia*) in the carrion sequence in Europe, but in the USA arrive with *Cynomyopsis* after *Lucilia* and *Cochliomyia* but before *Sarcophaga* (D. G. Hall, 1948; Rodriguez & Bass, 1983).

In Europe the species usually involved in forensic cases are *C. vicina* and *C. vomitoria* L., and the third instar larvae of these two species may be separated by the following couplet:

1 Posterior spiracles smaller (0.23–0.28 mm) and separated by a distance *equal to* or *greater* than the width of a single spiracle .... *vicina*
– Posterior spiracles larger (0.33–0.38 mm) and separated by a distance *less* than the width of a single spiracle .... *vomitoria*

Erzinçlioğlu (1985) provides keys to all the British *Calliphora* larvae; ecological notes on each species are provided below. It cannot be over-emphasized that, wherever possible, adults should be reared so that a specific identification can be established with certainty. For that purpose the adults of all the British species are keyed below. D. G. Hall (1948) keys the known larvae occurring in the Nearctic Region; O'Flynn & Moorehouse (1980) and Zumpt (1965) should be consulted for larvae of Australian species. The larva of the African *C. croceipalpis* has not been adequately described.

Key to British Calliphora *adults*
1 Basicosta of wing (Fig. 52, inset detail) usually orange, often brownish, but never black. Anterior thoracic spiracle orange (Fig. 49). Anterior two-thirds of jowls orange, posterior third black, but hairs black on both parts (very common, urban species) .... *vicina*
– Basicosta black. Anterior thoracic spiracle brownish 2
2 Hairs on lower part of jowls and along mouth-margin and on lower part of head orange in colour (very common rural species) .... *vomitoria*
– Hairs on jowls and head all black 3
3 Scutellum (Figs 49, 60) with three pairs of marginal bristles; spine at end of subcostal vein 2–3 times as long as the normal costal spines (Scotland, Inverness) .... *alpina*

– Scutellum with four or five distinct pairs of marginal bristles. Subcostal spine not distinct 4
4 Thorax viewed from behind evenly dusted with at most a few short narrow undusted streaks. Lower squamae dark brown up to the white embossed margin. Abdomen not tessellated (female often with jowls orange on anterior half) (Rare. Scotland: Sutherland, Hebrides, St Kilda. Western Ireland) .... *uralensis*
– Thorax viewed from behind with distinct undusted stripes. Lower squamae only slightly darkened. Abdomen distinctly tessellated 5
5 Facial dusting golden, male eyes separated above by width equal to that of third antennal segment (Scotland, northern England, uncommon) .... *subalpina*
– Facial dusting silvery; distance between male eyes above equal to 2.0–2.5 times width of the third antennal segment (Scotland, uncommon. Northern England) .... *loewi*

In general, *C. vicina* and *C. vomitoria* are the flies most frequently encountered in forensic cases involving cadavers in Europe.

Calliphora vicina Robineau-Desvoidy (= *C. erythrocephala* Meigen) is a common species widely distributed through the Holarctic region, which has followed man into South America, the Afrotropical Region (Mauritius and South Africa), northern India, Australia and New Zealand. In northern Europe it is a very common urban species closely associated with man.

Adults are attracted to faeces, decaying meat and fruit. The larvae (Figs 192–200) develop in carrion and the life history is shorter than that of *C. vomitoria* (see Table 5), due to less time being spent in the prepupal stage.

Reiter (1984) made a detailed laboratory study of the development of this species, under virtually field-like conditions of variable temperature, and recorded the following conclusions: the duration of egg stage increases at lower temperatures; the speed of larval growth is slower at lower temperatures; the maximal larval length is reached earlier at higher temperatures; the mean value of maximal length decreases with increasing temperature; larvae under all temperature conditions decrease in size after having reached their maximal length, the decrease in length being more rapid at higher temperatures; constant temperatures over 30°C lead to 'stunted forms' which do not pupate and subsequently die; constant temperatures under *c.* 16°C occurring *after* the peak of growth has been reached inhibit the readiness to pupate, causing the larvae to fall into a state of rest, which is interrupted only when the temperature is raised and the resumption of metamorphosis thus induced.

To facilitate a reconstruction of the larval age Reiter sets out the established growth data in the form of an *isomegalendiagram*, which facilitates temperature-fluctuations-related entomologic determination of the time of death with a maximum degree of accuracy (Figs 16, 45). If there is little fluctuation in temperature the age can be read off straight from the appropriate maggot length on the diagram. If a great fluctuation is probable then a time range can be read off between the maximum and minimum temperatures. Local meteorological stations will often supply air temperature records for the relevant period (see Methods and Techniques, p. 48).

Calliphora vicina is the commonest species found on human corpses, especially in urban situations (see Tables 1, 4, 5; Figs 44, 45 and Case histories p. 56) and is one of the few insect species to have whole monographs devoted to it – Lowne (1890–95) and more recently by Vinogradova (1984). Fabre (1913) is also well worth reading on this and other blowflies.

Calliphora vomitoria L. is another Holarctic species but it is essentially more rural in distribution then *C. vicina* and when the two occur together it is usually the less abundant. In Europe and Canada it is chiefly a forest species which ascends mountains and may be common around refuse in human settlements there. The larvae are carrion

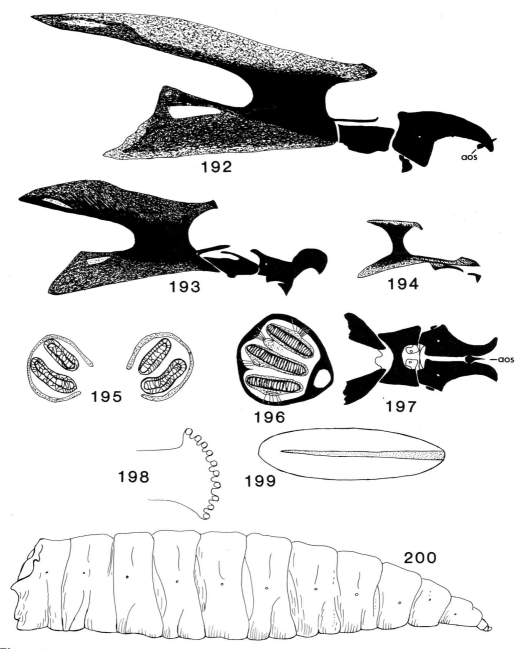

Figs 192–200 *Calliphora vicina* (Calliphoridae, Diptera), immature stages. 192, mouthparts, lateral, third instar larva (aos = accessory oral sclerite); 193, mouthparts, lateral, second instar larva; 194, mouthparts, lateral, first instar larva; 195, posterior spiracles, second instar larva; 196, posterior spiracles, third instar larva, 197, mouthparts, third instar larva, ventral; 198, anterior spiracle, third instar larva; 199, egg; 200, third instar larva, lateral.

feeders and the life history takes a little longer than that of *C. vicina* due to the longer time spent in the prepupal stage (see Table 5).

This species may be found on human corpses in rural situations (see Tables 1, 4, 5 and Case histories p. 56) instead of the urban *C. vicina*. However, *C. vomitoria* is commonly used for fishermen's maggots and may therefore be mechanically transported to areas where it does not normally occur.

Calliphora alpina Zetterstedt occurs in northern Europe, Greenland and northern Quebec. In Britain it was known from only two specimens but has recently been found in numbers at carrion bait in Cumbria and Durham (Dear, 1981).

Calliphora uralensis Villeneuve is an abundant species in the forest zones of the USSR, northern Europe, the mountains of central Europe (less commonly) and Greenland. In Britain it is uncommon and occurs in northern Scotland and western Ireland, and on Shetland, Harris, St Kilda and Fair Isle.

Larvae have been found in a wide range of food media, but are commonest in liquid excrement, especially in latrines and cesspools. Larvae do not seem to thrive in meat as well as *C. vicina* and are found less frequently.

Calliphora subalpina Ringdahl is another upland species in the forest zones of central and northern Europe, with a disjunct distribution. In the British Isles it is uncommon, but generally distributed between Inverness in Scotland and the southern Midlands of England.

Calliphora loewi Enderlein is another upland forest species in northern and central Europe. In Britain it is predominantly a northern Scottish species, but is also recorded from the border counties and as far south as Yorkshire, Cumbria and Durham.

Cynomya mortuorum (Linnaeus).

This is a moderately large (8–15 mm long) brilliant metallic, blue-green fly with the head mostly bright orange. It has a Holarctic distribution, but is commonest in the northern part of the region (Alaska, Greenland, Eurasia). In Britain it is a common northern English and Scottish species but it is scarce in central and southern England, the southern records being mostly coastal.

The larva (Figs 201–207) is a carrion feeder, but is not a myiasis producer. In Finland, Nuorteva (1977) found it attracted to raw fish baits and, to a lesser extent, human excrement, and gives a development time of 18–31 days in different sites during July and August at a mean temperature of 15°C. See also Erzinçlioğlu (1985) for details of larval structure.

Cynomyopsis cadaverina (Robineau-Desvoidy).

This is a moderately large metallic-blue fly with the head black above, and the face and front of the cheeks yellowish to reddish brown. It is found in the Nearctic region from Ungara Bay in northern Labrador to the Texas–Mexico border. It is commonest along the Canadian–USA border and has peaks of abundance in the spring and autumn. It usually arrives on carrion with *Calliphora*, following *Lucilia* and *Cochliomyia*, but preceding *Sarcophaga* (D. G. Hall, 1948; Rodriguez & Bass, 1983).

During warm spells adults may appear in midwinter, though it is believed that the species usually overwinters in the pupal stage. The larvae are normally found in carrion in an advanced state of decomposition, and a number of cases of myiasis in man and animals has been reported. Adults are also attracted to human excrement. In early spring and late autumn, in the warmer parts of its range, the adults of this species may enter houses in considerable numbers.

108 A MANUAL OF FORENSIC ENTOMOLOGY

Some larval structural details are illustrated by Tao (1927) and Teskey (1981).

Genus *Lucilia* Robineau-Desvoidy (= *Phaenicia* of American authors) (Figs 208–220, 304)

These are the familiar greenbottles, so called because of their brilliant metallic greenish coloration. The colour appears to vary with age. Newly emerged individuals are deep, bluish green with violet reflections, but with age this changes to an emerald green with

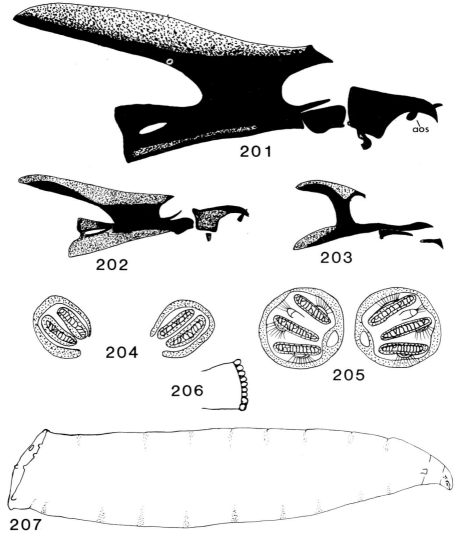

Figs 201–207 *Cynomya mortuorum* (Calliphoridae, Diptera), larva. 201, mouthparts, lateral, third instar (aos = accessory oral sclerite); 202, mouthparts, lateral, second instar; 203, mouthparts, lateral, first instar; 204, posterior spiracles, second instar; 205, posterior spiracles, third instar; 206, anterior spiracle of third instar; 207, third instar larva, lateral (partly after Schumann, 1962).

darker green reflections and, finally, to a dull, coppery green with reddish reflections (in the last type the wings are often worn and frayed). There is also inter-specific variation in the shades of green.

The adults (Fig. 304) are somewhat smaller (4.5–10.0 mm) and less bristly than *Calliphora*, but share similar habits and biology and some species are the well-known sheep maggot-flies (see Cragg, 1955; Norris, 1959, 1965a; Zumpt, 1965). There appears to be greater variation in the size of individual *Lucilia* than there is in *Calliphora*. This may reflect a greater sensitivity to the availability of food and adequacy of nutrition in the larval stages. Hanski (1976a) suggests that there may be genetic differences between populations of *Lucilia* and that small size could be an adaptation to environments whose competition for food is usually intense.

Lucilia sericata Meigen is the commonest attacker of sheep and is the species most likely to come to forensic notice. This and other important species are treated individually below. The genus is best represented in the Holarctic Region but occurs throughout the world.

As with *Calliphora*, adults especially males, are frequently found on flowers where they feed on the nectar. Unlike *Calliphora*, adults seldom come indoors and in the main prefer to be in the sunlight, though there is some inter-specific variation in this. Thus *L. illustris* Meigen is essentially an open woodland and meadow species, whilst *L. caesar* L. occurs in forests, appears to be more shade tolerant, and occurs further north than the former species. Most species of *Lucilia* are saprophagous and will breed in carrion and dung. Some may become facultative parasites and be involved in myiasis, one or two have become obligate parasites of Amphibia. *Lucilia* species are usually the first to arrive at carrion in sunlight.

Species of *Lucilia*, especially females, are not easy to identify and should be submitted to a specialist. Aubertin (1933) keys the world species, Schumann (1971) keys mid-European species (including larvae), and Zumpt (1956) keys African species. The following simplified key should enable males and some females of the British species to be identified.

Key to adults of British species of Lucilia

1 Basicosta (Fig. 52) yellow. Thorax with three posterior acrostichal bristles, i.e. strong bristles, one row each side of centre line, behind the thoracic suture (Fig. 60, ac) 2
– Basicosta brown or black 3

2 Tibia of middle leg with one antero-dorsal bristle. Male eyes separated by up to three times the width of the third antennal segment (very common species) .... *sericata*
– Tibia of middle leg with two antero-dorsal bristles (Fig. 51). Male eyes separated by up to twice the width of the third antennal segment (fairly common species) .... *richardsi*

3 Second abdominal segment, above, with strong bristles on the hind margin (parasite of frogs and toads) 4
– Second abdominal segment, above, without strong marginal bristles 5

4 Three posterior acrostichal bristles (uncommon) .... *silvarum*
– Two posterior acrostichal bristles (scarce) .... *bufonivora*

5 Eyes in male separated by the width of the third antennal segment. Antennal arista (Fig. 58) with up to ten hairs below (common) .... *illustris*
– Eyes in male separated by half or less than the width of the third antennal segment. Antennal arista with 12 or more hairs below 6

6 Male with large prominent bulbous genitalia with a ventral split in the tergite (very common species) .... *caesar*
– Male with inconspicuous genitalia (scarce species) .... *ampullacea*

Lucilia larvae should be reared through to adults wherever possible. The following simplified keys (adapted from Kano & Sato, 1952 and Zumpt, 1965) is incomplete, but

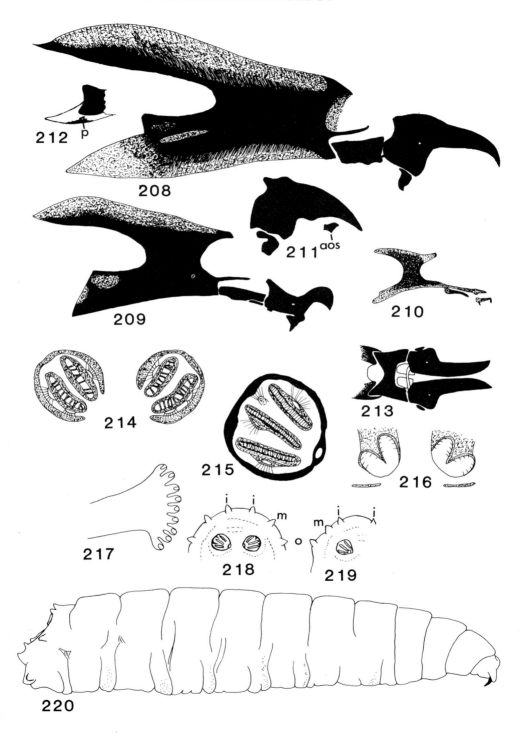

should facilitate the identification of a few common and important species. It should be used only where rearing is not practicable. Further critical research on the taxonomy of the larval stages is greatly needed.

Key to some third instar Lucilia *larvae found on carrion*

1 A small accessory oral sclerite present ventrally between the mouthhooks (smaller and shorter than in *Calliphora* or *Cynomya*; cf. Figs 192, 201, 211, aos) .... *ampullacea*
- No accessory oral sclerite (Fig. 208) 2
2 Mouthparts with a pigmented area below the posterior extremity of the ventral lobe (Fig. 212, p)
 .... *illustris* & *caesar*
- Mouthparts without a pigmented area below the posterior extremity of the ventral lobe 3
3 Upper margin of anal segment in end view (Fig. 218) with the inner tubercles (i) separated from each other by a distance approximately equal to the distance between the inner (i) and median (m) tubercles .... *sericata*
- Upper margin of anal segment with the inner tubercles separated from each other by a distance approximately equal to the distance between the inner (i) and outer (o) tubercles (Fig. 219, not British) .... *cuprina*

Lucilia sericata (Meigen) is widespread throughout the major zoogeographical regions, but is not yet cosmopolitan. It is very common in the temperate zone of the northern hemisphere, from whence it has followed man to many other parts of the world, and is now the dominant species in urban and suburban districts of Australia and Africa. It is present on St Helena and Tristan da Cunha but absent from Madagascar. In the New World it occurs from southern Canada to Argentina. It also occurs in the Oriental Region, but appears to be absent from the Pacific islands. In Britain it is a common species, generally distributed throughout England, Wales and the Scottish border counties, but has not been recorded further north than Ayrshire. *L. sericata* is unusually well adapted to competition with other carrion-feeding species and will certainly increase its range to other parts of the world, wherever there is sheep farming. It has been known as a causal agent of sheep strike since the Sixteenth Century. Adults are attracted by carrion, open wounds, the soiled or wet fleece of sheep, and faeces. The larvae can complete their development in all of these media.

Oviposition commences 5–9 days after emergence from the puparium. Each female produces 2 000 to 3 000 eggs in 9–10 batches over a period of about three weeks. *L. sericata* does not normally oviposit on carcases with surface temperatures below 30°C (Cragg, 1956). The eggs are yellow-white with a striated, faintly reticulated surface and about 1 mm long. In an English summer the incubation period is between ten and 52 hours on meat out of doors. On the back skin of sheep, under the wool, where the temperature is about 31°C, the larvae hatch within 8.75–10.25 hours. On a carcase the feeding period varies from five to 11 days, but at a constant temperature of 33°C this can be shortened to three days. At 39°C in the wound of a sheep, the first instar lasts about 12 hours and the second instar a further 12; after 43 hours at this temperature the third instar larvae will leave the wound and search for a pupation site. In summer conditions the prepupal period may last from three days to several weeks. In winter conditions in Europe the

Figs 208–220 *Lucilia* (Calliphoridae, Diptera), larvae. 208, *L. sericata*, mouthparts, third instar, lateral; 209, *L. sericata*, mouthparts, lateral, second instar; 210, *L. sericata*, mouthparts, first instar, lateral; 211, *L. ampullacea*, mandible, third instar (aos = accessory oral sclerite); 212, *L. illustris*, detail (to smaller scale) of ventral sclerite of mouthparts of third instar; 213, *L. sericata*, mandibles, third instar, ventral; 214, *L. sericata*, posterior spiracles, second instar; 215, *L. sericata*, posterior spiracle, third instar; 216, *L. sericata*, posterior spiracle, first instar; 217, *L. sericata*, anterior spiracle, third instar; 218, *L. sericata*, third instar larva, end view; 219, *L. cuprina*, ditto (i = inner tubercles, m = median tubercles); 220, *L. sericata*, third instar larva, lateral.

prepupa remains inactive in the soil until the temperatures reaches 7°C and pupation commences at 8–11°C. In a mild winter when the temperature remains above freezing point, breeding takes place throughout the year. The pupal stage lasts 4–7 days at 32°C, 6–7 days at 27°C and 18–24 days at 12–13°C. Development in a wound may, therefore, be completed within nine days of infestation. These data are from Zumpt (1965), after Davies (1934) and Ratcliffe (1935) working with living sheep in the UK (see also Povolny & Rozsypal, 1968). However, since *Lucilia* females will anticipate death when ovipositing in wounds of sickly hosts (or even in the absence of wounds, Davis, 1928; Hudson, 1914), these data could well have forensic significance in cases where there was a time-lapse between mortal injuries (especially stab wounds) and death. Rates of development in carrion are given in Tables 4, 5.

Lucilia cuprina (Wiedemann) was originally described from China. Its medical and veterinary importance remained unknown until the 1930s when Australian scientists realized that it had been confused with *L. sericata* and was in fact their most important sheep strike fly. A similar situation was then found to exist in Africa. In fact the species is widely distributed from the Mediterranean to the Oriental Region, throughout the Afrotropical Region (including Madagascar, Réunion and Mauritius), and the Australian, Nearctic and Neotropical Regions. *L. cuprina* can be distinguished from *L. sericata* by the bright metallic green coloration of the front femora, which are black or dark-bluish metallic in the latter species. Identification of tropical members of this genus is best referred to a specialist.

The life history of this species (mistakenly as *L. sericata*) has been studied by B. Smit (1931) in South Africa; here it breeds in carrion and produces nine or ten generations a year. However, no eggs were laid during the winter. A female will lay about 1 000 eggs, which take from eight hours to three days to hatch. The shortest feeding period for all three larval instars together was found to be as little as two days in optimum conditions, but up to three weeks in cooler or less favourable conditions. On leaving the carrion, larvae were found to vary considerably in the time they took to come to rest as prepupae, i.e. from a few hours to several weeks. Actual puparium formation also varied from a few hours to days or weeks. In the summer the pupal stage lasts seven days, but in winter it may be up to 115 days.

Larvae can successfully develop in necrotic tissues of animals or man. Hopkins (1944) records six cases of human myiasis in Uganda, including one from an arm burned by lightning and another from an ear injured by burns. Fiedler (1951) describes the pathogenesis of *L. cuprina* in sheep (see Zumpt, 1965 for partial English translation).

Lucilia caesar is restricted to the Palaearctic Region and records from other regions appear to be misidentifications. The species has also been confused with *L. illustris* and any records before the work of Séguy (1928) and Aubertin (1933) should be treated with caution. In Britain the species is common and generally distributed. In Finland, Nuorteva (1964) found considerable differences in the ecology of the two species. *L. illustris*, extending its range some 600 km further north in Finland, occurs about two weeks earlier in the season, is less synanthropic and is more heliophilic. However, he found the diurnal and annual fluctuations of the population density almost identical for the two species. Zumpt (1965) found no significant differences between larvae of *L. caesar* and *L. illustris*.

Lucilia illustris has been confused with *L. caesar*, as discussed above. It is very common in North America, Finland and Japan. According to Bishopp (1915, as *caesar*), in the USA the larval period lasts some two to five days at fairly high temperatures and puparia may be formed within 3–12 days from the hatching of the eggs. During periods of low temperature the third larval stage may last up to several weeks. The pupal stage lasts 5–16 days in fairly warm weather. Kano & Sato (1952) described the immature stages in some detail.

Lucilia ampullacea Villeneuve is another carrion species which has been confused with *L. caesar* in the past. It appears to be a common species in the Far East (including northern India), but is rarer and has a patchy distribution in Europe. In the British Isles it is an uncommon species though generally distributed throughout central and southern

England. It does not occur in the Afrotropical, Neotropical or Nearctic Regions, and Australian records are probably misidentifications.

Lucilia porphyrina is a common oriental species known from Sri Lanka (Ceylon), India, China, Malaya, Nepal and the Philippines. It develops in carcases, is frequently found in human dwellings and also causes myiasis in toads. It is closely related to *L. ampullacea* and can only be satisfactorily distinguished by examination of the male and female terminalia. It has also been confused with *L. papuensis* Macquart which occurs in the Oriental and Australian Regions.

Lucilia richardsi Collin is a European species and in the British Isles is fairly common and generally distributed through central and southern England.

Lucilia silvarum Meigen is a Holarctic species, recorded from Europe, North Africa, Eurasia, Japan and North America (from Canada south to Oklahoma). It has been confused with *L. bufonivora* and also independently recorded as a parasite of toads, but this needs more research.

L. bufonivora Moniez is an obligate parasite of toads and other amphibians and does not develop in carrion. The species is recorded from Europe, China and North America, but it is probable that the North American records are referable to *L. silvarum* or *L. elongata* (Shannon).

Other less known members of the genus may assume forensic importance in particular areas.

Phormia regina (Meigen).

This is a medium-sized fly (7–9 mm long) with a dark green to olive green body, a largely black head, and conspicuous orange hair on the anterior thoracic spiracle. It has a Holarctic distribution and is well known as a breeder in animal carcases. In the USA it is also recorded as a myiasis-producing fly (James, 1947), but Zumpt (1965) notes that it has not been so recorded in the Old World. In man it has been recorded in cases of dermal myiasis and is one of the species (see also *Protophormia terraenovae*) that has been used to clean wounds sustained by soldiers (Imms, 1939; Greenberg, 1971–73, vol. 2: 16).

The eggs are laid in masses and the larvae feed in large numbers in carcases. James (1947) reports that in the USA development is rapid, the total period from egg to adult ranging from ten to 25 days. Kamal (1958), also in the USA, gives a total development time from egg to adult of 10–12 days at 22°C and 50% RH (see Table 5 for details). James notes that cool weather favours development and that in the southern states adults are scarce in the hot months but can be found out of doors during the entire winter, at least as far north as the state of Iowa. It is a common sheep maggot in southwest USA. Denno & Cothran (1975) found that *P. regina* prefers bigger carcases and arrives later in the succession. The larva can be distinguished from *Protophormia terranovae* by the characters given in the key to Calliphorine genera.

Protophormia terraenovae (Robineau-Desvoidy).

This is a medium-sized (8–12 mm long), dark blue species with a greenish-blue abdomen and black legs. It has a Holarctic distribution and is very common in the cooler regions. It is most common in the spring and in North America it becomes more common than *Phormia regina*, especially at high elevations during the summer. Abundant in the Arctic, it has been found within 550 miles of the North Pole. In subarctic regions it is abundant in July. In Britain it is distributed widely and can be very common in wastes and marshes. It overwinters in the adult stages when the flies sometimes enter houses.

The larvae (Figs 221–225) develop in carrion and may be common in slaughter house yards. They also produce cases of myiasis in sheep, cattle and reindeer, but there seem to be no records from man. The larvae have been found to produce an antibiotic of possible relevance in wound therapy (Pavillard & Wright, 1957). Busvine (1980) notes that, unlike other blowflies, this species often pupates on the surface of the breeding medium unless it is very wet or exposed to bright light. Kamal (1958) gives a total development time of 10–13 days (see Table 5 for details).

Nuorteva (1977) reports a forensic case of Dr F. Mihályi in Hungary involving *P. terraenovae* in which the hatching time of the eggs (14–16 hours) was significant (see Case history no. 7, p. 58). He also reports a murder case in Finland in which the rate of development of *P. terraenovae* played an important role (see Case history no. 11, p. 60).

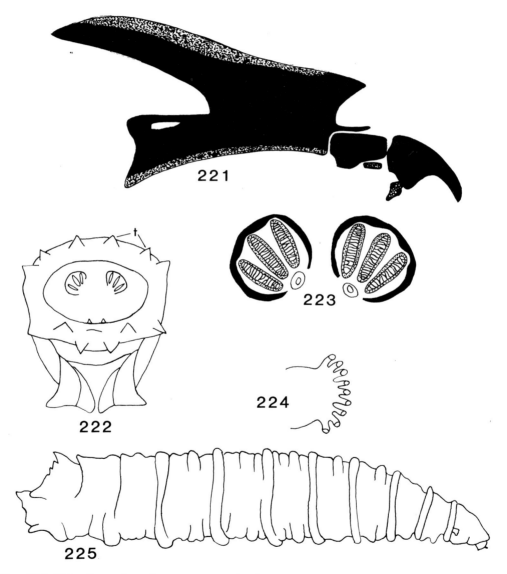

Figs 221–225 *Protophormia terraenovae* (Calliphoridae, Diptera), third instar larva. 221, mouthparts, lateral; 222, last segment, end view; 223, posterior spiracle; 224, anterior spiracle; 225, whole larva, lateral.

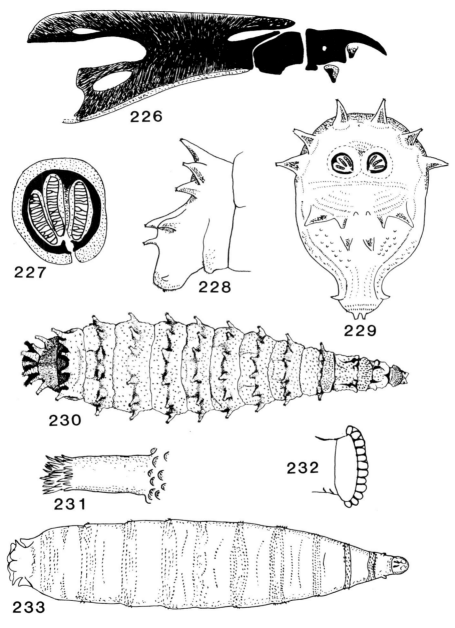

Figs 226–233 *Chrysomya* (Calliphoridae, Diptera), third instar larva. 226, *C. albiceps*, mouthparts, lateral; 227 *C. albiceps*, posterior spiracle; 228, *C. albiceps*, end segment, lateral; 229, *C. albiceps*, end view; 230, *C. albiceps*, whole larva, dorsal; 231, *C. albiceps*, detail of fleshy process; 232, *C. chloropyga*, anterior spiracle; 233, *C. chloropyga*, whole larva, dorsal.

Genus *Chrysomya* Robineau-Desvoidy (Figs 226–233, 303)

This is a common and abundant genus of the Old World tropics and subtropics, where it largely replaces the *Calliphora* and *Lucilia* of the Temperate Zone, and is the Old World equivalent of the New World *Cochliomyia*. The genus is included here as some species occur in the Palaearctic Region, including southern Europe, and in these days of rapid and easy travel it might have forensic involvements, especially as several species are successfully establishing themselves in new regions (e.g. Hanski, 1977a; Guimarães et al., 1978; Baumgartner & Greenberg, 1984).

The adults are large (c. 5–12 mm), thick-set flies of dark green or blue, sometimes metallic, coloration (Fig. 303). They are not easy to identify, but keys are provided below.

The immature stages are most likely to be encountered forensically, but are not described for all species. The following simplified key to larvae is based on Zumpt (1965) and Oldroyd & Smith (1973).

Key to some third instar Chrysomya *larvae*

1 Body with transverse rows of fleshy processes (Figs 230, 231) 2
- Body smooth with the only fleshy processes confined to the last segment (Fig. 233) 3

2 Mature larvae reaching a length of 18 mm. Body segments with a great number of long processes (Fig. 230); peritreme of posterior spiracle with a narrow opening and more or less distinctly forked at both ends (Fig. 227)
.... (Afrotropical & Neotropical) *albiceps* and (Oriental & Neotropical) *rufifacies*
- Mature larvae not more than 11 mm long. Body segments with fewer processes; peritreme of posterior spiracle with a broad opening, its ends not distincly forked ..... (Australian) *varipes*

3 Segments with belts of strongly developed spines. Anterior spiracles with 4–6 branches   *bezziana*
- Segments with belts of moderately developed spines. Anterior spiracles (Fig. 232) with 10–13 branches 4

4 Posterior spiracles closely approximated, separated by about one-fifth of the diameter of a peritreme .... *chloropyga*
- Posterior spiracles more widely separated, by one-third to one-half of the diameter of a peritreme .... *mallochi* and *megacephala*

Key to Chrysomya *adults associated with wound myiasis or carrion* (modified from Zumpt, 1965)

1 Small (up to 5 mm long); metallic bluish-black species; legs yellow with blackish stripes
.... (Australia) *varipes*
- Larger (6–12 mm long) species; legs coloured otherwise 2

2 Wings deeply infuscated at base and along anterior margin 3
- Wings clear, with only the base sometimes a little infuscated 4

3 Male with eyes touching above and with sharply demarcated enlarged upper facets; female with eyes separated above by a broad parallel-sided frons which, like the face, is yellow to orange. Body dark blue or bluish-green, abdomen with black transverse bands; legs uniformly dark
.... (Afrotropical, Oriental) *regalis* (= *marginalis*)
- Male with eyes touching above, but eye facets graduated in size from larger upper facets to slightly smaller lower facets; female with frons blackish and distinctly narrowed below near the antennal groove. Colour as preceding species .... (Afrotropical) *inclinata*

4 Anterior thoracic spiracle dark orange to blackish brown 5
- Anterior thoracic spiracle white to light yellow 7

5 Upper (thoracic) squama waxy white. Male eyes touching above, the upper facets slightly larger than the lower ones and graduated in size to them; female with eyes broadly separated and the frontal stripe almost parallel-sided; body metallic green or blue; legs dark
.... (Afrotropical, Oriental) *bezziana*

– Thoracic squama dirty grey to blackish brown. Male eyes touching; female with frontal stripe widened in middle. Colour as in preceding species 6
6 Male eyes with upper facets strongly enlarged and sharply demarcated from smaller ones on lower third; female with eyes broadly separated .... (widespread) **megacephala**
– Male eyes with upper facets distinctly enlarged, but not sharply demarcated from the smaller lower ones; female eyes broadly separated .... (Australasia) **mallochi**
7 No prostigmatic bristle. Body green to bluish; abdomen with dark transverse bands; legs dark; male eyes narrowly separated; female eyes broadly separated; cheeks of both sexes yellow to orange, only darkened posteriorly .... (Afrotropical, Oriental, Neotropical) **albiceps**
– Prostigmatic bristle present (below anterior spiracle) 8
8 Body metallic dark blue and green; front of thorax with an L-shaped mark facing outwards on each side; legs blackish. Eyes close together in male, broadly separated in female
.... (Afrotropical, Neotropical) **chloropyga**
– Body metallic bright green or bluish, without the distinctive L-shaped thoracic markings. 9
9 At least anterior part of the cheeks yellow or orange; male with inner and outer vertical bristles
.... (Oriental, Australasian, Neotropical) **rufifacies**
– Cheeks in both sexes black; male lacking outer vertical bristles
.... (Afrotropical, Neotropical) **putoria**

Chrysomya albiceps (Wiedemann) and *C. rufifacies* (Macquart) are treated together because their larvae are very similar and distinct from all other larval Calliphoridae (Fig. 230). *C. albiceps* extends from the southern Palaearctic Region (North Africa eastwards to north-west India), throughout Africa (including the Afrotropical islands of Aldabra, Cape Verde, Madagascar, Mauritius, Réunion, Rodriguez, ? St. Helena, Seychelles, Socotra) and has been introduced and is now established in Brazil (Guimarães *et al.*, 1978) and the Canary Islands (Hanski, 1977a). *C. rufifacies* replaces it in the Oriental and Australian regions.

Chrysomya albiceps is normally a carrion breeder and is frequently involved in secondary myiasis in sheep (following primary damage by *Lucilia cuprina* Wiedemann, at least in southern Africa), but cases in man are not known. Erzinçlioğlu & Whitcombe (1983) have established that it also breeds in sheep and goat dung.

The life history of *C. albiceps* has been studied in southern Africa by B. Smit (1931) and Cuthbertson (1933), and is summarized by Zumpt (1965). Eggs are laid on carcases in batches of 100–200, frequently among the eggs of other blowflies. Depending on the temperature, the eggs hatch in 24–36 hours and at first feed on the exudations of the decomposing flesh, but in the second and third instars may become predatory and even cannibalistic. M. Coe (1978) made the interesting observation that its larvae preyed upon the larvae of *Chrysomya marginalis* (Wiedemann) when both species were dispersing from an elephant carcase in Kenya (see Geographical location p. 28).

During the summer the larvae migrate from the carcase after about four days to search for pupation sites, the flies emerging about a week later. Temperature affects rate of development and Smit (*op. cit.*) found that in the Cape the feeding time could extend to 25 days but, in Zimbabwe (Southern Rhodesia), Cuthbertson found that in the dry season the feeding time lasted about a week and the pupal stage up to two weeks.

Patton (1922, as *C. albiceps*) and Roy & Siddons (1939) have studied the life history of *C. rufifacies* in India, while M. J. Mackerras (1933) and Norris (1959) have carried out similar studies in Australia. Cheong *et al.* (1973) report a forensic case involving *C. rufifacies* and *C. megacephala* in Malaysia. In Calcutta larvae hatched from the eggs after 8–12 hours, reached maturity in five days and the pupal stage lasted four days. At the lower temperatures in Canberra the life-cycle took 12–18 days.

I have seen larvae of *C. albiceps* from ancient Egyptian mummies.

Chrysomya varipes (Macquart) is known only from Australia where it breeds in carrion and is probably predaceous in the later larval stages.

Chrysomya chloropyga (Wiedemann) is a common blowfly in southern Africa where it is a primary and secondary cause of myiasis in sheep. It also occurs in cases of wound myiasis in cattle and has been recorded in human wound and intestinal myiasis (Zumpt, 1965). It has recently been found in Brazil (Guimarães *et al.*, 1978) and the Canary Islands (Hanski, 1977a). Lothe (1964) found this species on human corpses from medico-legal autopsies in Uganda. He found that the life cycle took about 17 days, but noted that oviposition could take place as late as three days after death. He concluded that when fully grown larvae were found, the body had been dead *at least* 3.5 days, and if empty puparia were found under the body, death had occurred *at least* 12 days previously.

The life history has been studied in detail by B. Smit (1931). The flies love sunlight and are attracted by carrion from a great distance. The 450 or so eggs are deposited in crevices in the carrion, very rarely on faeces, and hatch from 12 hours to three days later. The first instar lasts about 4–12 hours and the second instar about 36 hours. The third instar larva feeds for three days, the prepupal period lasts 1–2 days and the flies emerge about three days later. These are summer figures and in the cool season the time of development is extended so that the third instar may last four weeks, the prepupal stage three weeks and the fly may not emerge for a further 1.5 months. Zumpt (1965) gives the shortest time from egg to egg as two weeks.

Chrysomya bezziana Villeneuve is an obligate wound parasite in the Afrotropical and Oriental Regions and never breeds in carrion or decomposing organic matter.

Chrysomya mallochi Theowald is known only from New Guinea and northern Australia, where it is an important element in the carrion fauna and has been involved in cases of animal and human myiasis (Zumpt, 1965).

Chrysomya megacephala (Fabricius) is widely distributed over the Oriental and Australian regions and occurs in adjacent parts of the Palaearctic Region. It does not occur in most of mainland Africa, but is known from South Africa and the Afrotropical islands of Madagascar, Mauritius, Réunion and Rodriguez. It is also found on some Pacific islands and has been introduced and become established in Brazil (Guimarães *et al.*, 1978). Adults are common near human dwellings and thus likely to be encountered in forensic work in these regions. The flies are a great nuisance in open-air meat and fish markets, in slaughter houses, and around latrines and cess pools. They are also attracted to sweet foods and will enter homes to feed on them.

The biology of this species has been studied by Wijesundara (1957) in Sri Lanka (Ceylon). Approximately 220–320 eggs may be deposited on carcases, faeces, etc., usually in masses on the under surface. Development time from egg to adult takes eight and a half days at room temperature in Sri Lanka. The egg stages last 9–10 hours and the larval stages some 94–95 hours. The larvae are easily reared in the laboratory on pieces of meat and, at a constant temperature of 30°C, Wijesundara found that the life cycle could be completed in only seven and a quarter days, the egg stage being shortened by one hour and the larval stages by 5–6 hours. The duration of the first larval stage is 15–18 hours. In South Africa, Prins (1982) found the total life cycle to vary from 276 to 306 hours. Under laboratory conditions at 26°C Prins found the egg stage lasted 14 hours, the first larval instar 23 hours, the second instar 21 hours, and the third instar 60–72 hours, before they burrowed into the soil and pupated.

Bohart & Gressitt (1951) reared adult *C. megacephala* and found that they were strongly attracted to carrion, human excrement and sweets. They noted that old carrion, and faeces which had dried on the surface, had little attraction even when pungent. When flies were very numerous they formed large clusters on branches near their breeding material. This was particularly noticed in the vicinity of primitive outdoor toilets and during the early American occupation of Guam in World War II, when breeding in corpses was heavy. Orderly mass migrations of maggots from the corpses, all in the same direction, with a gradual thinning of numbers as they burrowed into the sand for pupation, are vividly described.

O'Flynn & Moorehouse (1980) have reared *C. latifrons* from a human cadaver in Australia.

Chrysomya regalis Robineau-Desvoidy (= *marginalis* (Wiedemann)) is widespread in the Afrotropical Region (including Madagascar and Socotra) and occurs in Pakistan. According to Prins (1982), it is a very common carrion breeder in South Africa (especially in March and April) and is a first-wave blowfly. By the time *C. albiceps* is attracted, large numbers of *C. regalis* have already hatched and by their vigorousness compete

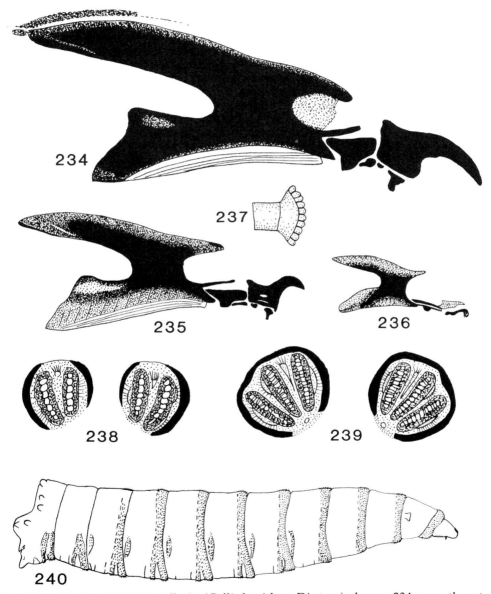

Figs 234–240 *Cochliomyia macellaria* (Calliphoridae, Diptera), larva. 234, mouthparts, third instar; 235, mouthparts, second instar; 236, mouthparts, first instar; 237, anterior spiracle, third instar; 238, posterior spiracle, second instar; 239, posterior spiracle, third instar; 240, third instar larva, lateral. (After Laake, Cushing & Parish, 1936.)

successfully both intra- and inter-specifically. However, *C. regalis* may be subordinate to *C. chloropyga*, which is also a first-wave blowfly. Prins gives a larval lifespan of 11 days during late summer in South Africa (6–9 days in *C. chloropyga*), and a pupal stage of nine days (5–8 days in *C. chloropyga*) during summer and some 14 days in spring. This species pupates beneath the carcase.

Other works giving developmental information and facilitating the identification of the larvae of *Chrysomya* species are Fan (1957) (for China); Ferrar (1979), Fuller (1932), Kitching (1976), Kitching & Voeten (1977), O'Flynn & Moorehouse (1980), O'Flynn (1983) (for Australia); Kano (1958), Ishijima (1967) (for Japan); and Prins (1982) (for South Africa).

Cochliomyia macellaria (F.) (Figs 234–240).

This New World species is called the secondary screw worm to distinguish it from the screw worm *Cochliomyia* (= *Callitroga*) *hominivorax* (Coquerel) (= *americana* Cushing & Patton), which is an obligate parasite and feeds in living tissue of man and animals and cannot exist in carrion. *C. macellaria* also occurs in cases of myiasis, but is a secondary invader and is primarily a scavenger; it may be very abundant in carrion in the New World.

The adult is metallic greenish blue with a yellow to reddish-orange face and three dark thoracic stripes. The species occurs in the Nearctic Region from Canada to the southern USA and in the Neotropical Region from Mexico to Panama. The adults prefer warm humid weather and are commonest during rainy periods. They are attracted to carrion, garbage and plants which emit an odour of carrion. The fly is extremely abundant in slaughterhouses and is common in outdoor markets in the tropics (Greenberg, 1971–73, vol. 1). It appears on carrion after *Lucilia* but before *Calliphora, Cynomyopsis* and *Sarcophaga* (D. G. Hall, 1948; Rodriguez & Bass, 1983). Denno & Cothran (1975) found that this species prefers larger carcases. A total of 1000 or more eggs are laid in loose masses of 40–250. Sometimes several females lay together, forming large masses of thousands of eggs. In favourable conditions the eggs hatch in four hours and the larvae (Figs 234–240) reach maturity in 6–20 days, when they migrate from the body to search for pupation sites in the soil. The total development time takes 9–39 days depending on temperature and humidity. Adults live from two to six weeks (James, 1947).

A full account of the morphology and biology of this species in comparison with the primary screw worm (*C. hominivorax*) is given by Laake *et al.* (1936).

The Nearctic *Paralucilia wheeleri* (Hough) has been confused with both species of *Cochliomyia* and may be found breeding in carcases but appears to be of little importance (Hall, 1948).

Fanniidae

(Figs 55, 241–246)

The Fanniidae are mainly Holarctic with a few species occurring in the Afrotropical, Oriental and Australian regions. The principal genus, *Fannia*, includes the lesser housefly (*Fannia canicularis* L.), which is the common domestic fly seen hovering beneath lamp-shades and which never seems to settle (in contrast to the larger *Musca domestica* L. which is more prone to settle). Males are usually found swarming beneath trees.

Fannia has no sharply angled (outer portion of the fourth longitudinal) vein as in *Musca domestica* and is further distinguished by wing vein 6 (anal) being short and vein 7 (axillary) sharply curved (Fig. 55). Males often have curious modifications of the legs.

Fannia canicularis, the lesser housefly, is a cosmopolitan species commonest in periods of warm weather but does, on the whole, seem to prefer cooler conditions than *Musca domestica,* and in Britain is the more common fly in houses until July. It is commoner in

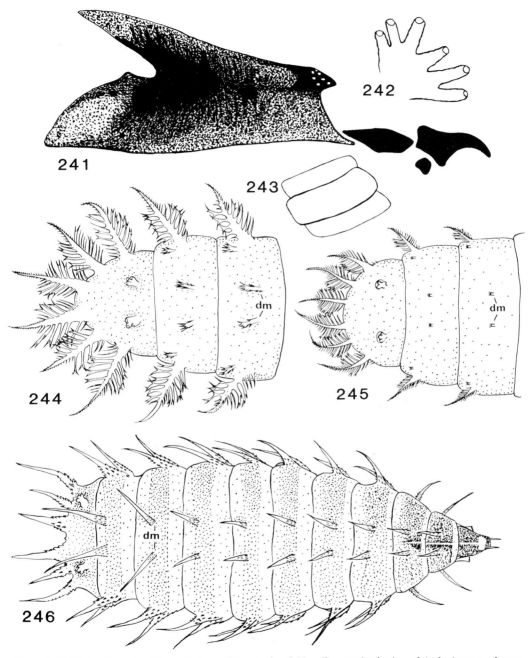

Figs 241–246 *Fannia* (Fanniidae, Diptera). 241, *F. canicularis*, third instar larva, mouthparts, lateral; 242, *F. canicularis*, anterior spiracle; 243, *F. canicularis*, egg (sculpture not shown); 244, *F. scalaris*, posterior end of third instar larva, dorsal view; 245, *F. manicata*, posterior end of third instar larva, dorsal view; 246, *F. canicularis*, third instar larva, dorsal.

rural areas, especially near poultry farms where the larvae breed in large accumulations of faeces. The flies are strongly attracted to urine and are common in toilets, privies, stables, pigsties etc. On a fox corpse, larvae of this species and of *F. manicata* (Meigen) were found about seven days after death and were present for some four weeks (K. G. V. Smith, 1975). In the USA, Wasti (1972) found larvae of *F. canicularis* on fowl carrion, 4–10 days after death, and Payne (1965) found *Fannia* larvae by the eighth day on pigs.

The eggs (Fig. 243) are about 1 mm long and banana-shaped with longitudinal flat lateral processes, an adaptation for floating in a semi-liquid medium. The larvae (Figs 244–246) are flattened and have characteristic lateral branched protuberances which also assist in floating. They live as scavengers in decaying organic matter including excrement. At 27°C (80°F) eggs hatch in 1–1.5 days. At 27°C (80°F) the larvae feed up and pupate in 8–10 days, the pupa lasts some 9–10 days and females will mate and oviposit in about 4–5 days. This gives a total egg-to-egg time of 22–27 days at this laboratory temperature.

Fannia scalaris F. has a cosmopolitan distribution and is essentially an outdoor species in contrast to *F. canicularis*. Its occurrence indoors usually indicates primitive conditions associated with lavatories and cesspits, and it is frequently called the latrine-fly. Optimum development occurs in semi-liquid masses of faeces, especially that of pigs but also of other animals including man.

These and other species of *Fannia* have been recorded in cases of urogenital myiasis and also on human carcases, usually after caseic fermentation.

The larvae most likely to be found in forensic investigations are illustrated. They have also been found in urine-soaked babies' napkins and cot-blankets (Teschner, 1961). Nuorteva (1974) recorded *F. canicularis* and *F. manicata* from a blood-stained shirt and the latter also occurred in dirty wet clothes in a washbasin (see Case history 18, p. 66).

Chillcott (1961) provides keys to the adults and larvae of Nearctic species of *Fannia* and Assis Fonseca (1968) keys adults of the British species. Lyneborg (1970) keys the larvae of 18 European *Fannia* species.

The *Fannia* larvae most likely to be involved in forensic cases may be identified as follows.

1 Dorsomedian processes (dm) long; the middle pair of processes on the last abdominal segment a little shorter than the inner and outer pairs and with shorter simpler lateral projections (Figs 246) .... *canicularis*
– Dorsomedian processes (dm) shorter; processes on last segment of equal size and with long, bifurcate lateral projections on the basal half (Figs 244, 245) 2
2 Dorsomedian processes with several projections (Fig. 244) .... *scalaris*
– Dorsomedian processes reduced to small button-like prominences (Fig. 245) .... *manicata*

Muscidae

(Figs 31, 53, 54, 247–290, 307–308)

The family Muscidae has a world-wide distribution and includes the housefly, face-flies, stable-flies, sweat-flies, etc. Some former subfamilies are now given family status, e.g. Fanniidae (which includes the lesser housefly), and are treated separately in the present work. Several species of Muscidae are of medical importance because of their relationship with man and his dwellings. These synanthropic species breed or settle on excrement, garbage, sewage, compost or carrion and may then become mechanical transmitters of disease when they settle, regurgitate or excrete on foodstuffs intended for human consumption. The synanthropic habit also ensures that certain Muscidae are likely to become involved in medico-legal cases, especially in domestic situations. The role of the few carrion-feeding Muscidae in the faunal succession is an important one as, apart from direct larval feeding on carrion, some species become predaceous in later instars on other carrion-feeders.

The few species usually involved in forensic investigations of cadavers are readily

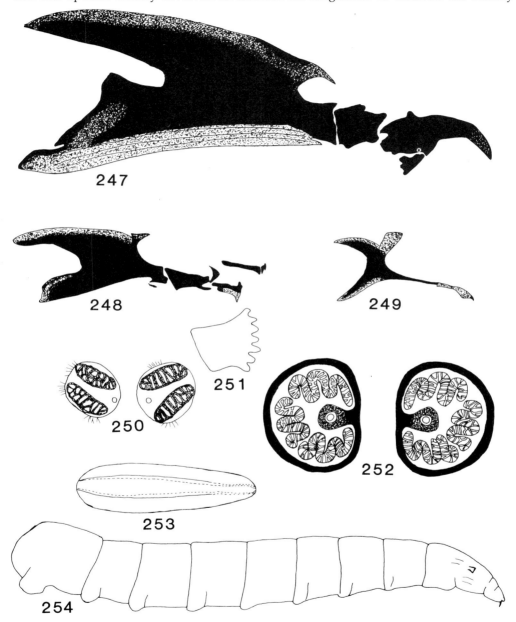

Figs 247–254 *Musca domestica* (Muscidae, Diptera). 247, mouthparts, third instar larva, lateral; 248, mouthparts, second instar larva, lateral; 249, mouthparts first instar larva, lateral; 250, posterior spiracles, second instar larva; 251, anterior spiracle, third instar larva; 252, posterior spiracles, third instar larva; 253, egg; 254, third instar larva, lateral.

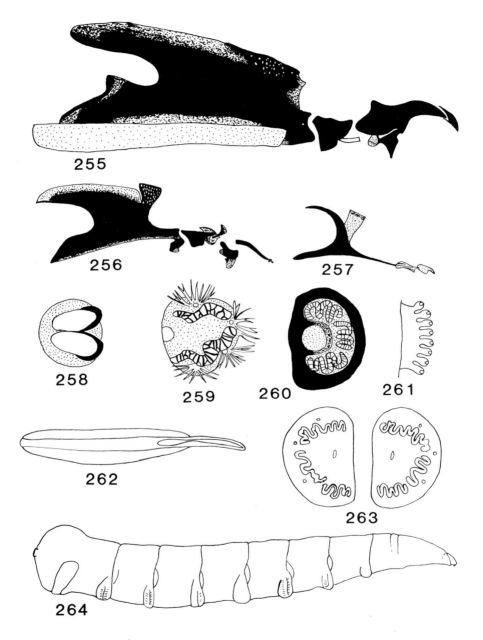

Figs 255–264 *Musca autumnalis* (Muscidae, Diptera). 255, mouthparts, third instar larva, lateral,; 256, mouthparts, second instar larva, lateral; 257, mouthparts, first instar larva, lateral; 258, posterior spiracle, first instar larva; 259, posterior spiracle, second instar larva; 260, posterior spiracle, third instar larva; 261, anterior spiracle, third instar larva; 262, egg; 263, spiracles, third instar larva (pigmentation not shown); 264, third instar larva, lateral.

distinguishable from each other and are briefly described below. Keys to adults for British members of the family are provided by Assis Fonseca (1968), and Skidmore (1985) treats the larvae on a world basis.

Occasionally, greenish metallic coloured species of Muscidae, which could be confused with the forensically important greenbottles (*Lucilia*, Calliphoridae) may occur in or near buildings or even on carrion. Two of these, *Dasyphora* and *Pyrellia*, may be distinguished from *Lucilia* at once because wing vein 4 is only bowed or curved toward vein 3 (as in *Muscina*, Fig. 54) and not sharply bent (see Figs 52, 304). The third 'false greenbottle' is *Orthellia*, which has the sharply bent fourth vein of *Lucilia* but, unlike that genus, the frons and jowls are also metallic green. These genera are more likely to be associated with excrement.

Genus *Musca* (Figs 53, 247–254, 255–264, 287, 312)

Musca domestica L. is the housefly, a truly synanthropic species that has followed man around the world. The adult fly (Fig. 312) is of medium size (6–7 mm), mouse-grey in colour and can be distinguished from other houseflies, except *M. autumnalis* DeGeer, by the sharply angled outer portion of the fourth long wing vein (Figs 53 a, 312).

It occurs commonly in houses where it will settle on food, refuse, faeces or on man himself. However, it is not a persistent sweat or mucus seeker. The adult will feed on almost anything with a moist surface and is particularly attracted to human faeces, manure heaps, liquefying garbage, meat and sweet foods. See also Levinson (1960).

In natural conditions, in rural areas, eggs are laid mainly on horse dung, but human, cow and poultry dung or material contaminated with excrement, decaying vegetable matter, garbage, decomposing foodstuffs, meat and carcases will all provide the emerging larvae with suitable nourishment. It is probable that oviposition on fresh corpses is rare and the involvement of the housefly in forensic medicine is more likely to occur when excrement in some form is present, e.g. either free, or if the gut contents are exposed. R. K. Chapman (1944) reports cases of *M. domestica* developing to adult on urine alone in infant bedding. The eggs may be laid singly or in small or large batches at the rate of 100–150 per day, up to a total of about 1000. The rate of development varies with temperature and may be summarized as follows. The immature stages are illustrated (Figs 247–254, 287).

Development of *Musca domestica* at different constant temperatures (after Busvine, 1980). Average durations of different stages (days).

	16°C 60°F	18°C 64°F	20°C 68°F	25°C 77°F	30°C 86°F	35°C 95°F	40°C 104°F
Egg	1.7	1.4	1.1	0.66	0.42	0.33	–
Larva	11–26	10–14	8–10	6.5	4.5	3.5	5.0
Pupa	18–23	12	9–10	6–7	4.5	4	4
Total	32	23	19	11	8	6	9

The housefly is one of the few species to have individual monographs devoted to it (Howard, 1912; Hewitt, 1914; West, 1951) and an individual bibliography (West & Peters 1973).

Musca autumnalis DeGeer (= *corvina* F.) is known as the face-fly, and is an important pest of cattle because of the irritation caused to the animals by the fly's habit of feeding in numbers on their body secretions, especially around the eyes and muzzle.

It resembles the housefly, but the male is easily distinguished by its bright orange-yellow abdomen with a black median stripe (Fig. 313). The female (Fig. 314) is distinguishable from *M. domestica* by the breadth of the frontal stripe, which is less than twice the width of an eye orbit in *M. autumnalis* and three or four times as wide in *M. domestica*.

The fly overwinters as an adult and has usually attracted attention in the English literature by its habit of aggregating in very large numbers indoors in late summer, especially in country areas. The immature stages (Figs 255–264) are spent in cattle droppings and it is only rarely found in carrion. When it does occur on a corpse it is usually in the first wave, when the body is fresh.

Genus *Ophyra* Robineau-Desvoidy (Figs 265–273, 286, 307)

Ophyra leucostoma (Wiedemann). This shining blue-black species (Fig. 307) is common throughout the Holarctic Region as far south-east as India and is frequently encountered around lavatories, slaughter houses, dung in poultry houses, etc. The males hover in the air like some Syrphidae (hover-flies).

The larvae develop in excrement including human faeces, offal and carcases. In the second and third instars the larvae become predaceous and frequently attack other larvae living in the same medium, including the larvae of *Musca domestica* and other Muscidae. Puparia (Fig. 286) have been found about three feet deep in the soil beneath the pabulum in March. Adults emerge when the mean temperature beneath the surface of the ground reaches 50°F (10°C) and are abundant from June to October, being most plentiful in August (Graham-Smith, 1916).

In human corpses *Ophyra* usually appears during the period of ammoniacal fermentation, some four to eight months after death. Mégnin (1894) found *Ophyra* in myriads in corpses about a year old. If the body has not been exposed to the open air for that period of time these may be the only maggots or puparia present and the time of their arrival may be affected. On a burnt full term foetus found an open ground there was no evidence of blowfly feeding, but larvae of *Ophyra* and *Fannia* were present.

The egg (Fig. 272) is about 1 mm long by 0.3 mm wide, slightly curved with a pair of longitudinal ridges on the ventral surface, and with hexagonal surface sculpture.

The mature larva (Fig. 273) is about 12.5 mm long by 2 mm wide. It is white with a thick hard skin. Diagnostic features of the three larval instars are illustrated (Figs 265–271). The puparium (Fig. 286) has conspicuous prothoracic horns (Fig. 269).

Ophyra capensis (Wiedemann) has similar habits to *O. leucostoma* and occurs widely in the Palaearctic and Afrotropical Regions and into India, with records from the USA and Chile. It prefers a warmer environment than *O. leucostoma* (natural or artificial). It has been reared in large numbers from the rotting head of a rorqual whale from the Hebrides and also found swarming near a bone factory (Assis Fonseca, 1968). I have seen it from human bodies that have been kept indoors for several months, where blowflies have not had access (see Case history 19, p. 66). This species has appeared in the literature as *O. anthrax* Meigen and *O. cadaverina* Mégnin and this synonymy has been resolved by Pont & Matile (1980). Cuthbertson (1935) has illustrated details of the larva of *O. capensis* (as *anthrax*). I have been unable to find any significant differences between the larvae of the two species.

The immature stages of the British *Ophyra* species need further study and their precise role in the carrion fauna clarified. It seems that *O. capensis* is probably a later arrival than *O. leucostoma*. Adults should be reared out to establish an identification, using the following key.

Key to adult British Ophyra *species* (modified from Assis Fonseca, 1968)

1 Lower of the two squamae (lobes behind wing base Fig. 52) darkened with dark brown border and fringe; anterior bristles on top of thorax well developed; male with hind tibiae distinctly curved and densely hairy .... *leucostoma*

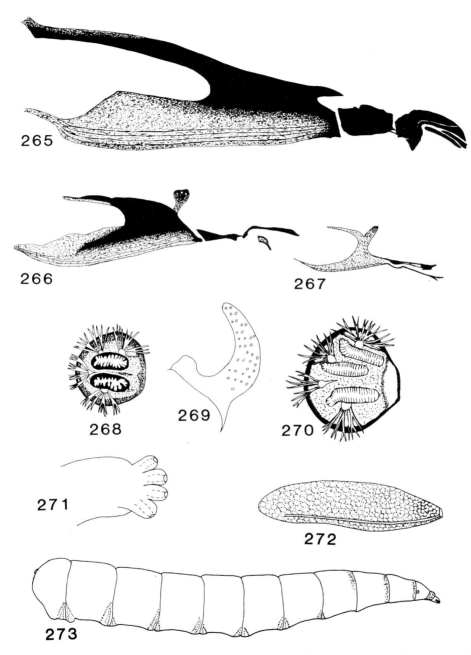

Figs 265–273 *Ophyra leucostoma* (Muscidae, Diptera). 265, mouthparts, third instar larva, lateral; 266, mouthparts, second instar; 267, mouthparts, first instar larva, 268, posterior spiracle, second instar larva; 269, spiracular horn of puparium (see also Fig. 286); 270, posterior spiracle, third instar larva; 271, anterior spiracle, third instar larva; 272, egg; 273, third instar larva, lateral.

– Lower squama white with white border and fringe; bristles on top of thorax hardly distinguishable from the hairs; male hind tibiae straight .... *capensis*

Ishijima (1967) describes the larvae of *O. leucostoma*, *O. chalcogaster* (Wiedemann) and *O. nigra* Wiedemann, which have been found on dead animals in Japan (*O. capensis* has not been found there).

Hydrotaea dentipes (F.) (Figs 274–278, 288, 308)

A common, dark-brown fly of similar size (8 mm long) to the housefly, but without obvious bends in the wing veins. The male is much darker than the female, has smoky wings (which are clear in the female), and a ventral tooth-like spur and a tubercle distally (Fig. 308b) on the front femur.

Adults frequent faeces and decaying fruit and are common around latrines and offal pits. It has an Holarctic distribution and is associated with man under primitive conditions.

The larvae develop in accumulated compact and drying human faeces for preference (Greenberg, 1971–73, vol. 1), but are also found in decaying vegetables and dead animals (Ishijima, 1967). In the third instar the larvae become predatory; Graham-Smith (1916) claims that the species is the most important of those with carnivorous larvae and says 'The enormous destruction of the larvae of common flies wrought by the larvae of *H. dentipes* alone has not hitherto been taken into account.' Nuorteva (1977) records oviposition by this species in Finland on a human corpse partly covered in melting snow and (1974) from a bloody shirt (see Case history 18, p. 66). Keilin (1917) and Zimin (1948) describe the third larval stage in some detail. Lobanov (1968) keys the larvae of eight species of *Hydrotaea* (see also Skidmore, 1985).

The related common species, *H. occulta* (Meigen), also occurs on carrion and is distinguished from *H. dentipes* by differences in the front tibia in the male (cf. Fig. 308a and b), and in both sexes the eyes are densely hairy (the hairs are longer in the male). This species has been reared from dead snails and sheep (Beaver, 1969) and I found adults attracted to a fox corpse a few days after death. Payne & King (1972) reared it as a scavenger in pig corpses in water.

The most important British Muscidae which have carnivorous larvae were placed in order of abundance of the adults by Graham-Smith (1916) as follows:

Hydrotaea dentipes (F.)
Muscina stabulans (Fallén)
Graphomya maculata (Scopoli)
Polietes albolineata (Fallén)
Phaonia erratica (Fallén)
Azelia macquarti (Staeger)

Some of these occur on carrion and the adults may be identified using the keys of Assis Fonseca (1968).

Genus *Muscina* Robineau-Desvoidy (Figs 54, 279–285, 289, 290, 315)

Flies of this genus resemble *Musca*, but have the apex of the wing vein gently curved rather than sharply bent (Figs 54, 315) and, consequently, the tips of veins 3 and 4 are wider apart. They are greyish flies with a tessellated abdomen and a reddish tip to the scutellum. All are common and generally distributed and may occur on animal or human corpses. An interesting feature of *Muscina* species is that, unlike most carrion flies, they are not deterred from oviposition when a corpse is covered by a thin layer of soil

(Nuorteva, 1977; see Case histories 9 (p. 59) and 17 (p. 65)). They will lay their eggs on the soil surface and the young larvae reach the carrion by moving down through the soil (see also Heleomyzidae, p. 81).

They seem to have a preference for human faeces and are not commonly found on carrion. In the literature they are usually associated with fresh corpses, though Thomson (1937) states 'carrion in the later stages, when it is already consumed by larvae of *Calliphora, Lucilia* and *Hydrotaea*, is specially attractive, not only to *Muscina pabulorum* Fallén but also to *M. stabulans* Fallén, but in the case of the latter species I have never been able to observe or induce oviposition on this substance'. Keilin (1919) records cocoon formation by the larvae inside which the puparia are formed (Figs 289, 290). This could make the 'on site' search for puparia in the soil difficult in forensic cases (see Procedure on Site, p. 39).

Muscina stabulans Fallén is a cosmopolitan species. It is distinguished from other members of the genus by having all tibiae and the apical part of the mid and hind femora

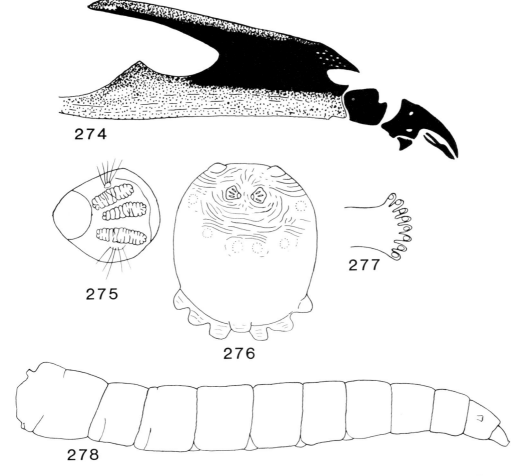

Figs 274–278 *Hydrotaea dentipes* (Muscidae, Diptera), third instar larva. 274, mouthparts, lateral; 275, posterior spiracle; 276, end view of last segment; 277, anterior spiracle; 278, whole larva, lateral.

yellow. Adults (Fig. 315) are usually found near human habitation in the country, especially in the open around stables, poultry houses, on the walls of outside lavatories or wherever there is decaying organic matter. Adults are frequently found in houses. The larvae (Figs 279–285) are found in a wide range of media including rotting fungi, fruits, broken eggs, excrement and carrion. They appear to be saprophagous in their first instar and become predaceous in later instars. In birds' nests the larvae will attack nestlings, causing their death, and larvae have also been found in the dead bodies of other insects.

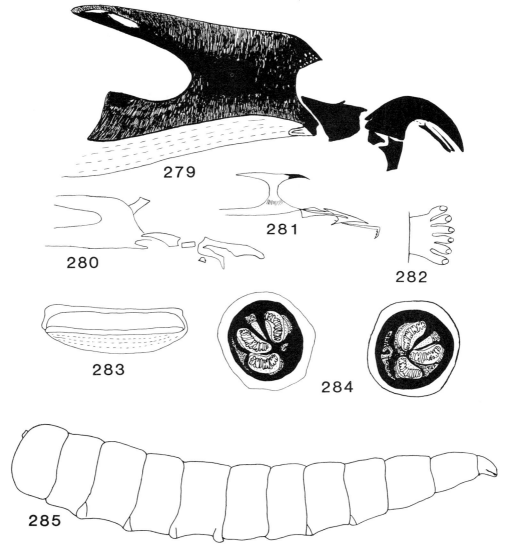

Figs 279–285 *Muscina stabulans* (Muscidae, Diptera). 279, mouthparts, third instar larva; 280, mouthparts, second instar larva (after Séguy, 1923*a*); 281, mouthparts, first instar larva (after Séguy, 1923*a*); 282, anterior spiracle, third instar larva; 283, egg; 284, posterior spiracle, third instar larva; 285, third instar larva, lateral.

Mégnin (1894) found *M. stabulans* to be one of the species present in exhumations carried out in cemeteries and which became active after the body was placed in the coffin. Nuorteva (1974, Case history 18, p. 66) reared this species from larvae on a bloodstained shirt. See also Portschinsky (1913).

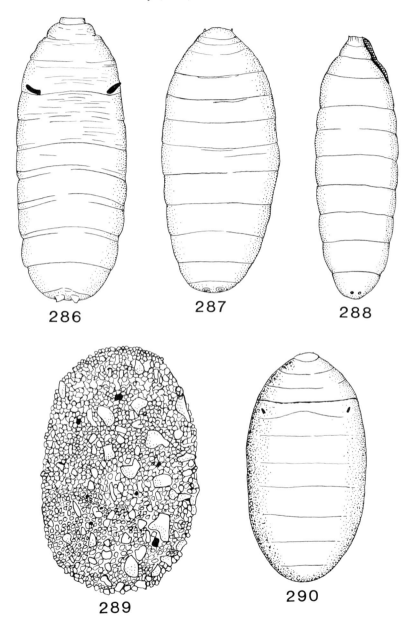

Figs 286–290 Puparia of Muscidae (Diptera). 286, *Ophyra capensis*; 287, *Musca domestica*; 288, *Hydrotaea dentipes* (empty); 289, *Muscina pabulorum*, cocoon; 290, *M. pabulorum*, puparium.

Muscina pabulorum Fallén has black legs and reddish to yellowish palpi. Like *M. assimilis* (see below) its distribution is Holarctic, and it has similar habits and larval habitats, but the adults rarely enter houses. It is more usually found in carrion and is purely saprophagous in the first and second instars and, according to Thomson (1937), will not attack living prey, even in the absence of its normal food. In the third instar, however, it becomes a facultative carnivore and will attack and devour other carrion-frequenting larvae. Keilin (1917) records it as a predator in silk-moth larvae and on the migratory larvae of the fly *Sciara militaris* (Sciaridae). As quoted above, Thomson (1937) found *M. pabulorum* readily attracted to carrion and while sometimes females laid only a few eggs, at other times 60–150 were laid. He found that the eggs took from 36–38 hours to hatch, although as long as 66 hours has been recorded. Unfortunately he gives no temperature or humidity data, but goes on to record that on a diet of dead meat the larvae could be fully grown and ready to 'pupate' (form a puparium) seven days after hatching. The pupal stage i.e. from puparium formation to emergence of the fly, is recorded as lasting about three weeks.

Muscina assimilis Fallén has black palpi and is frequently seen basking on tree trunks or bare patches of ground in early spring. It has similar habits and larval habitats to *M. stabulans* and *M. pabulorum*. Keilin (1917) gives detailed descriptions of the immature stages which are usually found in fungi. Nuorteva reared it from fish bait and the species has been recorded by Teskey & Turnbull (1979) from prehistoric graves on an archaeological site (2000–2500 years old) in Canada.

All three species of *Muscina* have also been reared from wasps' nests.

Figs 291–295 Adult Diptera (flies). 291, *Trichocera* sp. (Trichoceridae); 292, *Psychoda alternata* (Psychodidae); 293, *Conicera* (Phoridae, the coffin-fly); 294, *Megaselia scalaris* (Phoridae); 295, *Eristalis tenax* (Syrphidae), vs = vena spuria or false vein. (Fig. 291 McAlpine et al. 1981, Fig. 292 and 295 from Stakelberg, 1956.)

Figs 296–301 Adult Diptera (flies), acalyptrate families, 296, *Sepsis cynipsea* (Sepsidae); 297, *Madiza glabra* (Milichiidae); 298, *Dryomyza anilis* (Dryomyzidae); 299, *Piophila casei* (Piophilidae); 300, *Drosophila funebris* (Drosophilidae), a = arista, cb = costal breaks, vib = vibrissa; 301, *Centrophlebomyia furcata* (Piophilidae) (after Séguy, 1950). (Figs 296–299 from Stakelberg, 1956).

Figs 302–306 Adult Diptera. Calliphoridae (blowflies) and Sarcophagidae (flesh-flies). 302, *Calliphora vicina* (bluebottle); 303, *Chrysomya megacephala*; 304, *Lucilia sericata* (greenbottle); 305, *Protophormia terraenovae* (302–305 are Calliphoridae, blowflies); 306, *Sarcophaga carnaria* (flesh-fly, Sarcophagidae). (From Stakelberg, 1956.)

Figs 307–311 Adult Diptera (flies). Muscidae, Fanniidae and Scathophagidae, 307, *Ophyra leucostoma* (Muscidae); 308, *Hydrotaea dentipes* (Muscidae), a, *H. occulta*, male front leg, b, *H. dentipes*, male front leg; 309, *Fannia canicularis* (Fanniidae, the lesser housefly); 310, *Scathophaga stercoraria* (Scathophagidae, the common dung-fly); 311, *Copromyza equina* (Sphaeroceridae). (From Stakelberg, 1956 except for inserted details on Fig. 308.)

Figs 312–315 Adult Diptera, Muscidae. 312, *Musca domestica* (the housefly); 313, *Musca autumnalis*, male; 314, *M. autumnalis*, female; 315, *Muscina stabulans*. (From Stakelberg, 1956.)

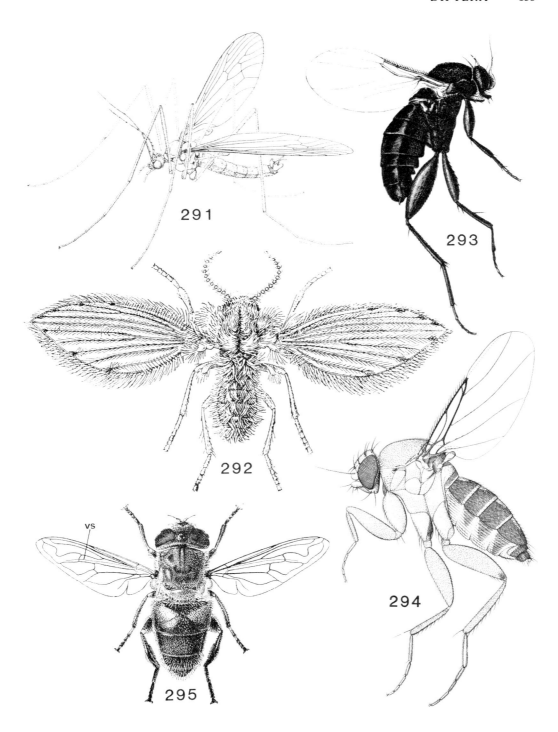

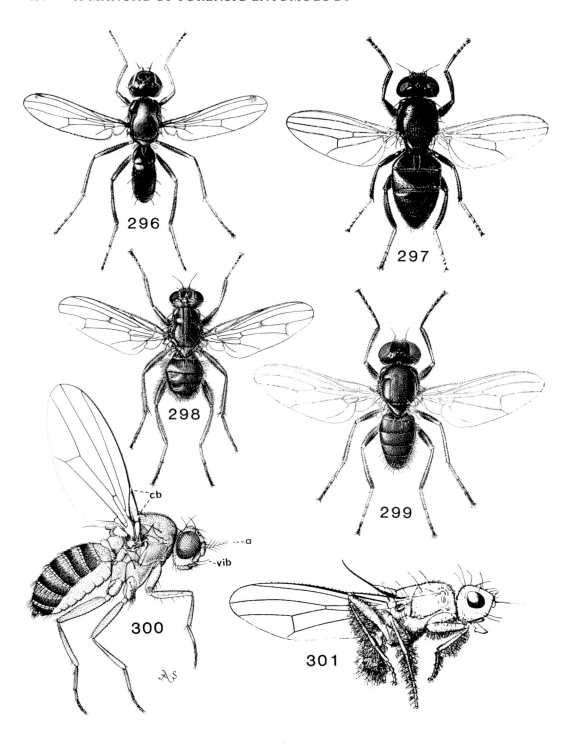

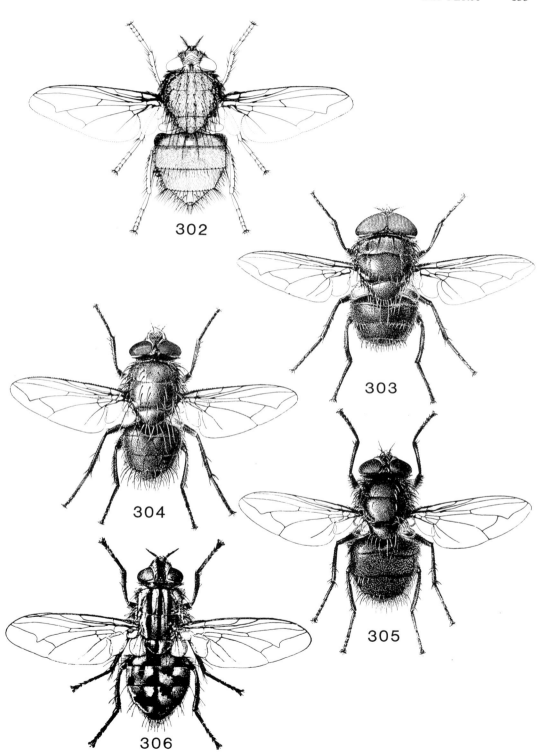

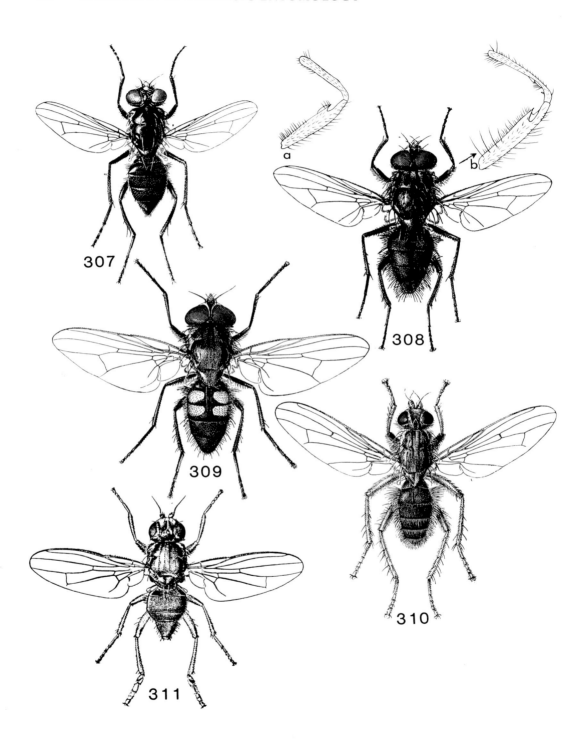

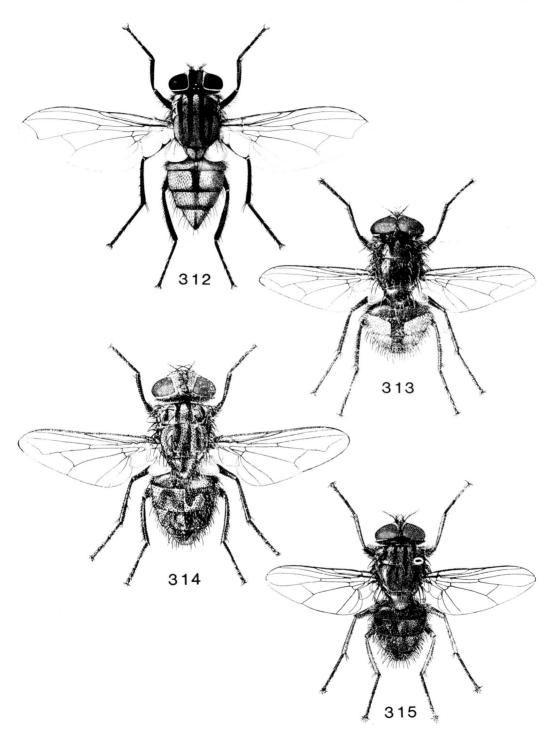

Coleoptera

Beetles and larvae
(Figs 316–357)

The Coleoptera, or beetles have the fore-wings hardened into elytra or wing cases, which usually cover the abdomen, giving the insect an armour-plated appearance. They are tough insects and vary enormously in size, from 0.5 mm to 150 mm in length. This is the largest order of insects with over 275 000 known species, of which over 4 000 have been recorded in Britain.

Most beetles can fly, but they are usually found on the ground, in soil, under bark, on vegetation, among debris, etc. Many are carnivorous and some are aquatic.

Beetle larvae are of varied appearance. Some possess thoracic legs and are very active, while others are grub-like, fleshy and relatively inactive.

The beetle fauna has probably received more attention in the literature than any other insect group occurring on carrion, though its true assessment as an indicator of forensic value has been largely overlooked. Morley (1907) provides an early assessment of the beetle fauna of carrion in Britain. Over a period of ten years he found 113 species, of which he regarded 37 as genuine flesh eaters. Other local British studies were made by Kaufmann (1937, 1941), Elton (1966) and Easton (1966). Easton in particular was anxious to redress the balance with regard to the forensic importance of beetles and in his detailed study of a fox corpse, over a period of 12 months, he found 1 967 individual beetles representing 87 species in ten families. The Staphylinidae (rove-beetles) were most numerous (1 054 individuals in 60 species). Some of his other statistics are cited under families below. As a keen amateur coleopterist as well as a police surgeon, Dr Easton's findings are particularly valuable. He compared (Easton & Smith, 1970) the species of beetles on the fox with those present on a human corpse, where he counted 98 beetles representing 40 species in six families. He found that several beetles restricted to the first five months on the fox were also present on human cadavers, e.g. the Staphylinidae *Proteinus brachypterus* (F.) *Anthobium unicolor* (Marsham), *Atheta fungivora* (Thomson), *A. ravilla* (Erichson) (= *angusticollis* (Thomson), *A. cadaverina* (Brisout) and the leiodid *Catops tristis* (Panzer). On the other hand, he found that several beetles visiting the fox only from the fifth month onwards were absent from the human, e.g. Clambidae, Ptiliidae, Histeridae, Nitidulidae and Cryptophagidae. However, he pointed out that the absence of the commonest species found on the fox, i.e. *Atheta aquatica* (Thomson), together with the absence of *Autalia* spp., is more probably explained by differences of environment. Nevertheless, these findings suggest fruitful avenues for more detailed comparative studies.

While many beetles are specifically associated with carrion, the majority are probably predators and only a few are true carrion feeders. The feeding habits of adults and larvae may also differ in a particular species and are discussed more fully under each family below. Crowson (1981) gives an interesting discussion on beetles in carrion and points out that 'competition with Diptera (as well as with vertebrate scavengers) must have had an important influence on the evolution of carrion beetles'. The greater mobility of adult flies facilitates their reaching a cadaver before the beetles, and their generally faster rate of larval growth enables Diptera to complete their development before the food source is

---

Figs 316–327 Coleoptera (beetles) larvae found on carrion. 316, *Nicrophorus* (Silphidae); 317, *Silpha* (Silphidae); 318, *Hister cadaverinus* (Histeridae); 319, *Catops picipes* (Leiodidae); 320, *Rhizophagus* (Rhizophagidae); 321, *Atomaria ruficornis* (Cryptophagidae); 322, *Cantharis* (Cantharidae); 323, *Atheta sordida* (Staphylinidae); 324, *Harpalus* (Carabidae); 325, *Cercyon* (Hydrophilidae); 326, *Philonthus* (Staphylinidae); 327, *Cryptopleurum* or *Megasternum* (Hydrophilidae) (after Bøving & Henriksen, 1938).

COLEOPTERA 139

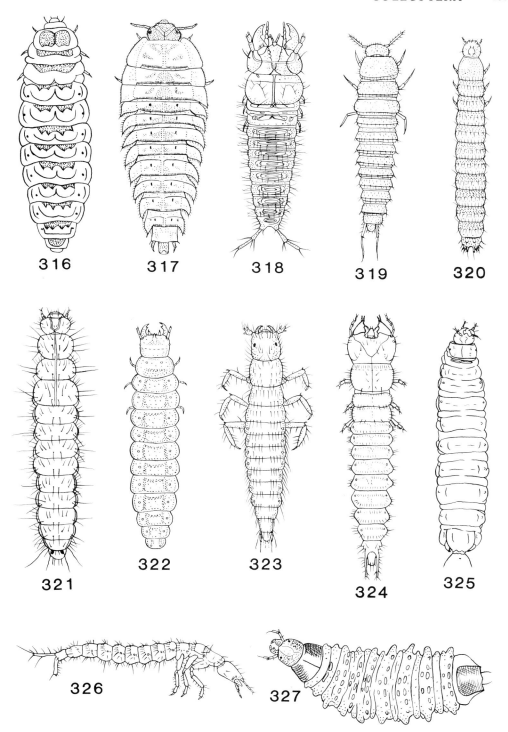

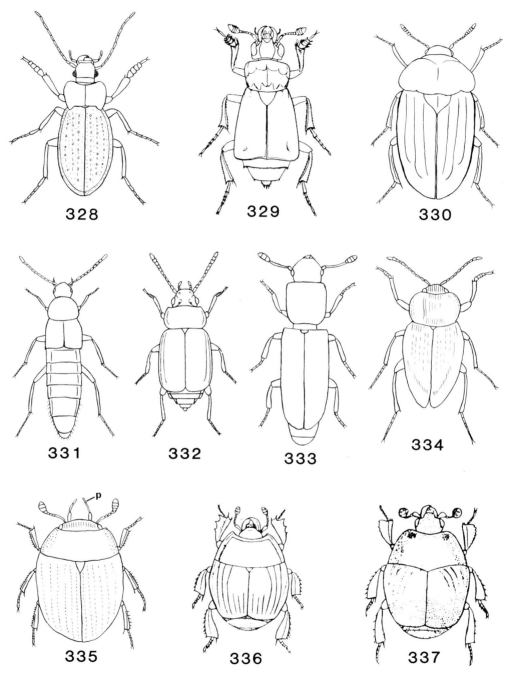

Figs 328–337 Coleoptera (beetles) adults found on carrion. 328, *Carabus* (Carabidae); 329, *Nicrophorus* (Silphidae); 330, *Silpha* (Silphidae); 331, *Oxypoda* (Staphylinidae); 332, *Anthobium* (Staphylinidae); 333, *Rhizophagus* (Rhizophagidae); 334, *Catops* (Leiodidae); 335; *Cercyon* (Hydrophilidae); 336, *Hister* (Histeridae); 337, *Saprinus* (Histeridae).

exhausted. Crowson points out another advantage possessed by Calliphorinae (blue- and greenbottle) larvae: their large numbers produce considerable amounts of free ammonia which is very toxic to carrion beetles. The beetles' competitive advantages are the generally better developed sensory and locomotor equipment of their larvae and the burrowing ability of their adults, features which enable them to select more favourable feeding areas and habits. Many carrion beetles avoid competition with blowflies by occurring on carcases at a later, drier stage rather than on the earlier, wetter corpses

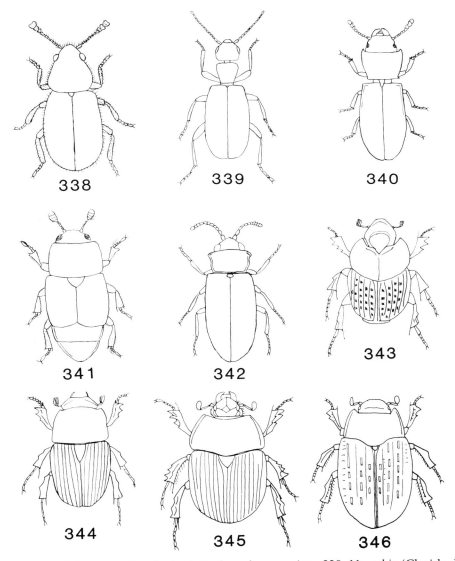

Figs 338–346 Coleoptera (beetles) adults found on carrion. 338, *Necrobia* (Cleridae); 339, *Anthicus* (Anthicidae); 340, *Tenebroides* (Trogossitidae); 341, *Carpophilus* (Nitidulidae); 342, *Cryptophagus* (Cryptophagidae); 343, *Onthophagus* (Scarabaeidae); 344, *Aphodius* (Scarabaeidae); 345, *Geotrupes* (Geotrupidae); 346, *Trox* (Trogidae).

favoured by flies. Completely dry corpses attract a distinct beetle fauna of Dermestidae, Tenebrionidae and Ptinidae (Figs 347–357).

In *Nicrophorus* (Silphidae), Springett (1968) describes a very subtle device, which has evidently evolved in competition with blowflies, involving a symbiotic relationship with mites (Acari) of the genus *Poecilochirus*. Adult mites are present, usually in numbers, on adult *Nicrophorus*. When the beetle finds a cadaver the mites move off and seek out fly eggs upon which they feed. Thus if the beetles arrive before the fly eggs have hatched they stand a much better chance of survival (see also availability of Food and Competition, pp. 34, 165).

The precise nature of the attraction of beetles to carrion has been investigated by Shubeck (1968) in the USA. His studies suggest that Silphidae, at least, are not as efficient in detecting carrion as had been formerly supposed. He found that there is a linear relationship between the distance at release and the rate of return to carrion (by *Oiceoptoma* (= *Silpha*) *noveboracensis* Forster) when release occurred at distances from five to 75 metres. This appears to be due to random wandering rather than orientation to carrion odours. He found a significant increase in the ability of the beetles to return to carrion below a distance of two metres and that the periphery of odour perception seems to be about one metre from carrion. He also makes the interesting comment that he could himself detect such carrion at about 25 metres distance! He further found that at a distance of five metres *S. noveboracensis* was 14 times more likely to return to carrion than *Nicrophorus*. Dethier (1947), working in New England, found a much greater overall return of Silphidae than in Shubeck's New Jersey study and the latter suggests that the effect of wind may account for the differences. Haskell (1966) states that 'while Dethier (1947) ascribed carcass finding ... to a klinokinetic mechanism, his description of the process fits better the idea of odour released anemotaxis'. Dethier also placed his carrion bait five feet above the ground, whereas Shubeck's was at ground level and probably gave more realistic results. It is also possible that different populations of carrion beetles may exhibit different behaviour patterns. Clearly, there is a large field open for research here and many questions need fuller investigation, e.g. full assessment of wind effects – does it initiate and affect the duration of flight, does the scent of carrion itself initiate flight and does this response depend on the physiological state of the beetles, detailed behaviour of beetles on carrion, precise niches occupied, precise nature of food and method of feeding and detailed life histories. Such studies would certainly establish that some beetles are as useful as blowflies in certain forensic cases.

There is no up-to-date general work for the identification of all British beetles to species. Joy (1932) is very useful if used carefully and in conjunction with modern check lists (because the classification and nomenclature is dated). The series *Die Käfer Mitteleuropas* (Freude *et al.*, 1965–84) is also useful. Several families are individually covered in the '*Handbooks for the Identification of British Insects*' published by the Royal Entomological Society of London. The introductory Handbook by Crowson (1956) gives a key to families and a similar introduction to beetle larvae is in preparation (by Dr Jane Marshall). Another useful book is Linssen *Beetles of the British Isles* (1959), and Hinton & Corbet's (1975) keys to stored products species cover several families found on carrion. Other specialist keys are cited in Kerrich *et al.* (1978) for Britain and northern Europe, and Hollis (1980) for the World. Crowson (1981) should be consulted for general biological information and Klausnitzer (1978) for a general account of beetle larvae which facilitates identification. The observations of Fabre (1918*a*, 1918*b*, 1919, 1928) are also well worth reading for detail.

The families usually associated with carrion are briefly discussed and illustrated below, although there is little data from human cadavers and much more work needs to be done before the potential forensic importance of beetles can be assessed. Where identification keys are not specifically cited under a family it may be assumed that there are none and reference should be made to the general works cited above.

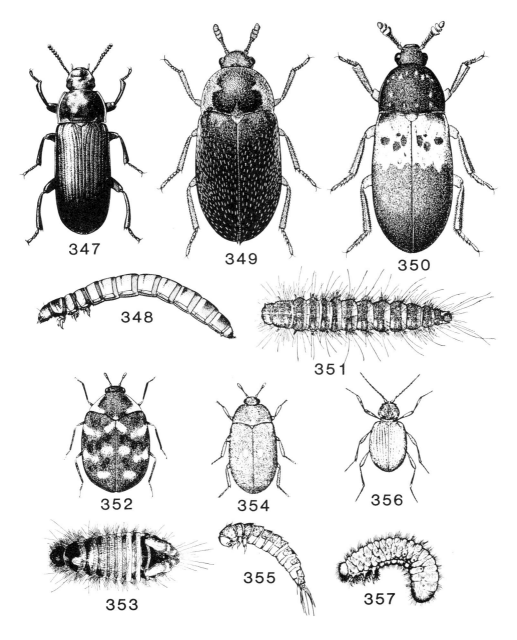

Figs 347–357 Coleoptera (beetles) found on dry corpses. 347, *Tenebrio molitor* (Tenebrionidae); 348, *T. molitor* larva (mealworm); 349, *Dermestes maculatus*; 350, *Dermestes lardarius*; 351, *D. lardarius* larva; 352, *Anthrenus verbasci* (Dermestidae); 353, *A. verbasci* larva; 354, *Attagenus pellio* (Dermestidae). 355, *A. pellio* larva; 356, *Ptinus tectus* (Ptinidae); 357, *P. tectus* larva.

Carabidae

(Figs 32, 328)

These are the well-known ground beetles of brown, black or sometimes metallic blue or green coloration which vary in size from 1 mm to 8 cm. They have a small head armed with strong jaws, long slender antennae and long legs clearly adapted for running.

Members of this family may be present on carrion as predators, though their importance as such is minor compared with the Silphidae, Staphylinidae ahd Histeridae. They are more likely to be seen at night. In his study of pig cadavers, Payne (1970) found adults of 14 species, but no larvae. On a dead fox Easton (1966) found one species, but Smith (1975) found none. R. F. Chapman & Sankey (1955) found *Harpalus rufipes* (DeGeer) and *Pterostichus niger* (Schaller) on rabbit corpses. Nabaglo (1973) found nine species on rodent carcases in Poland. Lindroth (1974) keys the British species.

Hydrophilidae

(Figs 325, 327, 335)

Although the name of this family suggests an aquatic life-style, in fact, many species are not found in water though the larvae occur in at least moist situations. The adults may be distinguished at once by their long palpi (Fig. 335, p), which are longer than the antennae. The terrestrial species are usually associated with mud, dung or rotting vegetable matter. Precise data on their feeding habits are not available, but several species occur on carrion.

Cercyon (Figs 325, 335). The larvae of this genus (see Bøving & Henriksen, 1938) live on land, in rotting vegetable matter near water (including under decaying seaweed on the seashore). They are predaceous, probably on dipterous larvae, in these habitats. Smith (1975) found an adult of *C. lateralis* (Marsham) on a fox corpse on 27 September 1972, some 31 days after death, where it was probably predaceous on the smaller Diptera larvae then present (Fanniidae, Stratiomyidae). Payne (1970) found four species of *Cercyon* in his studies of pig carrion in the USA. They were found on fresh pigs, both on land and floating on water. On the pigs on land they remained concealed beneath the carcases, where they probably frequently escaped notice due to their small size. Payne made no observations on their feeding habits. Several *Cercyon* species have also been recorded from carrion by Howden (1950), Moore (1955), R. F. Chapman & Sankey (1955) and Reed (1958).

Cryptopleurum, Megasternum (Fig. 327). Bøving & Henriksen (1938) describe the larva of what is probably one of these two genera from earth between daisy plants sent from Ireland to the USA. Both Easton (1966) and Smith (1975) found *Megasternum obscurum* (Marsham) and *Cryptopleurum minutum* (F.) on fox corpses, the former in May–June, some 5–6 months after death (both species) and the latter in September, ten days after death (*Megasternum*) and in November, two and a half months after death (*Cryptopleurum*).

Easton (*in* Easton & Smith, 1970) found *Cercyon terminatus* (Marsham), *C. analis* Paykull and *Megasternum obscurum* (Marsham) on 26 October 1969 by sieving surface soil and leaf litter beneath a human corpse on the North Downs, Surrey, some 17 days after death.

Balfour-Browne (1958) keys the British species.

Silphidae

(Figs 316, 317, 329, 330)

This family is very variable in size and shape and there have been conflicting opinions on the silphid feeding habits. Some workers suggest that the beetles feed both on carrion and the associated dipterous maggots. Clark (1895) suggested that dung or carrion beetles fell into three groups: feeding purely on the medium; feeding on the medium and to a limited extent on the maggots therein; and those feeding exclusively on maggots. Silphids were included in the last two categories. Steele (1927), working with *Silpha* and *Nicrophorus*, showed that some adult silphids preferred maggots and would die if limited to a carrion diet. Abbot (1937) found that silphid adults were partly predaceous. Dorsey (1940) found that the adults of only one of six species of *Silpha* he studied would feed on maggots and none in the larval stage. Illingworth (1927) reported that *Nicrophorus* was among the chief agents in reducing the number of dipterous larvae on a cat carcass he studied in southern California. Payne & King (1970), in their study of pig carrion in the USA, found that adults of *Necrophila (= Silpha) americana* L. and *Necrodes surinamensis* (F.) fed only on maggots, but the immature silphids appeared after most of the Diptera larvae had left the dried remains of the pig carcases and it was assumed that they fed on carrion. Heymons *et al.* (1926) have reared the larvae of *Silpha obscura* L. entirely on flesh.

Easton (*in* Easton & Smith 1970) found 13 *Necrodes littoralis* (L.) on the body of a man which had been lying on the North Downs for 17 days in October 1969. One of these held a dipterous maggot in its jaws and Easton suggests that maggots were probably the true source of attraction to the beetles (see also under Staphylinidae, below).

Leiodidae
(Figs 319, 334)

The genus *Catops* used to be placed in the Silphidae but is now regarded as a leiodid. They are small (less than 5 mm long), brown to black beetles, with their upper surface covered in a fine down.

Catops species are frequently found on or under carcases. Payne (1970) found as many as 50 individuals of *Catops* on a single piglet carcase. Easton (1966) found seven species on a fox corpse, some immediately after death, and Smith (1975) found *Catops tristis* (Panzer) beneath a fox corpse from one to two months after death, when the Diptera maggots were gone. Nabaglo (1973) found eight species of *Catops* on rodent carcases in Poland. Easton (*in* Easton & Smith 1970) records *C. tristis* from a human cadaver.

Staphylinidae
(Figs 323, 326, 331, 332)

This family is distinguished by the very short elytra or wing cases, but this does not reflect on the powers of flight of its members as a pair of efficient wings is tucked beneath. These beetles are probably the commonest predators found on cadavers, both as larvae and adults. Some species of the subfamily Aleocharinae have larvae which are ectoparasitic upon pupae of certain Diptera and may, therefore, also be present on a cadaver. The larvae of Diptera are the main prey of Staphylinidae but other insect larvae occurring on a corpse will also be taken. These beetles arrive and begin exploring a body within a few days of death and, during the bloated stage, they are present as long as there is insect activity on the corpse. Payne & King (1970) found over 50 species on pigs, and report that they were active during both the day and night. They found that on exposed carcases the immature stages appeared too late to feed on maggots, but were present on buried pigs when maggots were in abundance. They also note that adult staphylinids preyed heavily on dipterous larvae as the latter migrated from the corpse, even pursuing them into the soil.

Easton (1966) found 60 species on a fox corpse over a period of 12 months, and Smith (1975) found 11 species on another fox corpse from late August to November. R. K. Chapman & Sankey (1955) found six species on a rabbit corpse and report that the commonest beetle of all was *Aleochara curtula* (Goeze), a parasite of dipterous pupae.

They found that *A. curtula* first appeared on the third day after death and reached its peak on the seventh day, after which numbers fell rapidly (this was in July 1951). Nabaglo (1973) recorded 21 species from rodent carcases in Poland.

Easton (*in* Easton & Smith 1970) records two forensic cases involving staphylinid beetles. In the first the finding of 30 examples of *Anthobium atrocephalum* Gyllenhal (Fig. 332) in the wrappings of the corpse of a new-born child suggested that the body had been abandoned for a longer time than a pathologist's estimate had suggested. A longer period (of nine days) was later established on the subsequent confession of the mother. In another case involving the dumping of the body of a man, the presence of only one specimen of *Oxypoda lividipennis* Mannerheim (Fig. 331) suggested that the body had been on site for a much shorter period than the pathologist had suggested, and again this was confirmed by other factors. The same author records *Creophilus maxillosus* (L.) and 27 other species from the body of a man which had been lying on the North Downs for some 17 days in October 1969 (see also under Silphidae, p. 144 and case 4, p. 56).

The scattered literature on Staphylinidae is listed in Kerrich *et al.* (1978) and Hollis (1980). Kasule (1968) provides keys to genera for larvae (Figs 323, 326).

Histeridae

(Figs 318, 336–337)

Like the Staphylinidae, members of this family have short elytra though they are usually longer in histerids, so that only two abdominal segments are visible, and they are more squat in shape.

This family of beetles occurs wherever there is decay and putrefaction. In the USA, Payne & King (1970) found them on pigs during the bloated, decay, and early parts of the dry stage. They found that they were usually concealed beneath the carcase during daylight, became very active at night, and fed exclusively on dipterous larvae. Holdaway (1930) regarded histerids as the significant factor in the reduction of dipterous larvae. In Australia, Fuller (1934) found both adult and larval histerids feeding on blowfly larvae and puparia. Howden (1950) reported histerids feeding on maggots under both laboratory and field conditions in the USA. In Finland, Nuorteva (1970) confirmed that histerids cause a remarkable decrease in the numbers of developing blowflies, arriving several days later than the ovipositing blowflies and eating their eggs and larvae. He found that *Hister striola* Sahlberg, *Saprinus aeneus* (F.), *S. planiusculus* (Motschulsky) (= *cuspidatus* Ihssen) and *S. semistriatus* (Scriba) feed on the eggs and other stages of carrion flies (e.g. *Calliphora, Lucilia, Cynomya* and *Ophyra*), and *Hister abbreviatus* F. feeds on the eggs of *Musca autumnalis* DeGeer.

Saprinus and *Dendrophilus* occur on dead animals and on air-dried and smoked foods, where they prey on the larvae of *Dermestes* (Dermestidae), (Hinton, 1945).

Mégnin (1894) regarded histerids (*Hister*, Figs 318, 336 and *Saprinus*, Fig. 337) as occurring on human corpses after the remaining body fluids had been absorbed, about a year after death, but clearly they appear much earlier than this.

Halstead (1963) and Hinton (1945a, 1945c) will be found useful for identification.

Cleridae

(Fig. 338)

Beetles of this family are small and hairy, especially on the thorax. They are predaceous (probably on maggots) as adults and larvae, and occur on flowers, on dead or dying trees infested with borers, and on bones and skins of dead animals. Clausen (1940) states that members of the genus *Necrobia* have food and feeding habits different from the rest of the family. Clark (1895) states that *Necrobia* probably feeds on maggots but, in the USA,

Payne & King (1970) found that *Necrobia rufipes* (DeGeer) (Fig. 338) fed only on carrion and was found on the dried bones of 'treed' and 'grounded' pig baits.

Anthicidae
(Fig. 339)

These are small (3.0–3.5 mm) beetles, generally looking somewhat like Carabidae, although some are ant-like in appearance. Fuller (1934) found *Anthicus 'hoeferi* Kerg'. (error for *hesperi* King) in association with carrion in Australia. Payne & King (1970) record four species from 'insect-open' and 'treed' carrion (pigs) in the advanced decay and dry stages in the USA.

Dermestidae
(Figs 349–354)

These are small- (3.5–10.0 mm) to moderately-sized beetles densely covered with short hairs or scales which are often conspicuously coloured. They feed on a wide variety of dried animal matter, especially dried fish, hides, skins, furs, woollens, bird and rodent nests and some attack vegetable products. The larvae are hairy and sometimes called 'woolly bears'. The hairs have urticating properties and care should be taken when handling the larvae not to inhale hairs or get them in the eye.

Dermestes (Figs 349–351) are elongate beetles some 0.75–1.0 cm long. Females lay up to 150 eggs from which small hairy larvae hatch within about three weeks. The larval stages last from 5–15 weeks, depending on temperature and type of food available. The pupal stage lasts from two weeks to two months and beetles that emerge towards the end of the year overwinter in the pupal cell and emerge the following spring.

Anthrenus (Figs 352–353) (carpet- or museum-beetles) are more oval beetles of variegated coloration and about 3.5 mm long. The female beetle lays up to 100 eggs during a period of about two weeks, in spring and early summer. The eggs hatch in 2–4 weeks and the larval stage lasts for about a year. The fully grown larva becomes inactive and pupates within its last larval skin and the adult emerges from the pupa in 10–30 days.

Attagenus pellio (L.) (Figs 354–355) (the two-spotted carpet beetle) is an oval, black insect, some 6.5 mm long, with a small white spot on the middle of each elytron. From late April to August the female lays up to 150 eggs. The rate of development varies greatly with temperature, but as a rule there is one generation per year.

Dermestids can be very numerous feeding on skin, sinews and bone. Fuller (1934) collected 466 larvae from one sheep's head and Payne & King (1970) found 50 adults on one dried piglet carcase.

Dermestes first appear during Mégnin's third wave, when the fats are rancid, after some 3–6 months. *Dermestes*, *Anthrenus* and *Attagenus* also occur when the corpse is completely dry and when clothes moths (Tineidae) may also be present.

Keys to Dermestidae are provided by Hinton (1945b, 1945c) and Hinton & Corbet (1975). See also Strong (1981).

Nitidulidae
(Fig. 341)

These are small (2–4 mm) obovate or oblong beetles, dark brown in colour, sometimes with a yellowish spot at the apex of the somewhat shortened elytra. This is a very large family of over 2 000 species which are mostly sap-feeders on trees or on the juices of fruits. Many live on flowers and fungi, but some occur on carrion and a few are predaceous.

Payne & King (1970) found nine species on pig carcases during the dry stages, in company with Dermestidae. They noted that the nitidulids preferred moister skin than the dermestids.

Hinton & Corbet (1975) key some *Carpophilus* species (Fig. 341).

Rhizophagidae
(Figs 320, 333)

These are small narrow beetles about 2.0–2.5 mm long, often found under bark. They have been recorded from various dead animals (Howden, 1950), including dead fish (T. J. Walker, 1957). Payne & King (1970) found three species of *Monotoma* on pig carcases during the late stages of decay.

This family is particularly associated with buried human corpses, where *Rhizophagus parallelocollis* Gyllenhal is placed in Mégnin's fourth wave, some two years after death.

Peacock (1977) provides keys to the British species.

Ptinidae
(Figs 356–357)

These small (2–4 mm) beetles are commonly referred to as 'spider-beetles' because of their resemblance to small golden brown spiders (from which they are distinguishable at once by their long antennae and, of course, in possessing three, not four, pairs of legs). They commonly feed on grain, flour, dried fruit and other dry stored products. The female lays about 100 eggs over a period of 3–4 weeks and these hatch in about two weeks at summer temperatures. The larva (Fig. 357) is white and feeds up in about 40 days, when it makes a cocoon covered with the material on which it has been feeding. In a further 20–30 days the adult emerges. These beetles feed on dried carrion and on human corpses in Mégnin's eighth wave, some three years or more after death.

Adults may be identified with the keys in Hinton & Corbet (1975) and larvae are treated by D. W. Hall & Howe (1953).

Tenebrionidae
(Figs 347, 348)

These are brownish beetles, either dull (*Tenebrio obscurus* F.) or shining (*T. molitor* L.), about 13–25 mm long. They are commonly associated with birds' nests, but are also pests on stored foodstuffs, especially meal and flour, and the larvae are the well-known 'mealworms' (Fig. 348). Normally the life-cycle takes about a year, but it can extend to two, the larval period lasting throughout the winter. Tenebrionidae are found, along with Ptinidae, on human corpses in Mégnin's eight wave, some three years or more after death.

Scarabaeidae
(Figs 343, 344)

This is a very large family (over 19 000 world species, mostly tropical) of which 80 or so species are found in Britain and are commonly called 'chafers' and dung-beetles. They are very variable in shape and size and live in rotten wood, roots of plants, decaying vegetable matter, dung and some in carrion. The superfamily, Scarabaeoidea, also includes the two families, Geotrupidae and Trogidae, which may be distinguished at once among the carrion fauna by their antennal 'club', which consists of a number of lamellae or thin plates, and their fore legs which are conspicuously modified for digging. Many dung-beetles construct tunnels under the carcases.

Payne & King (1970) found 14 species of Scarabaeidae, mostly of the genera *Onthophagus* (Fig. 343) and *Aphodius* (Fig. 344), on exposed pig carcases. They note that none was found on 'tree' carcases, nor on those in water or buried. Pessoa & Lane (1941) give a full account of species of medico-legal interest in Brazil.

Keys to the British Scarabaeidae can be found in Britton (1956).

Geotrupidae
(Fig. 345)

This is another family of dung- or 'dor-' beetles belonging to the superfamily Scarabaeoidea, which may be found on excrement or carrion. They are robust beetles, and construct tunnels beneath the carcases which they stock with carrion for their larvae. Clark (1895) stated that *Geotrupes* feeds entirely on carrion and dung and this is confirmed by later workers.

Nuorteva (1977) records *Geotrupes stercorosus* (Scriba) on a human corpse in Finland (along with histerids and staphylinids) partly covered by moss and by the branches of a rotten tree (see Case history 9, p. 59).

For identification of the British species see Britton (1956).

Trogidae
(Fig. 346)

This is another family of the Scarabaeoidea which is usually associated with the nests of small mammals and birds, fungi, carcases, decayed fish and occasionally dung. Clark (1895), Abbott (1937), Spector (1943) and Payne & King (1970) have all reported Trogidae as carrion feeders. Howden (1950) and Reed (1958) record them as being attracted only to dry carcases, but Payne & King (1970) found eight species on pigs during the advanced stages of decay, as well as on dry remains.

The three British species of *Trox* may be identified using Britton (1956).

Other Coleoptera

Some families recorded from carrion, but which were probably more specifically feeding on fungi or fungal hyphae present on the corpses include the Corylophidae, Ptiliidae, Erotylidae, Mycetophagidae, Lathridiidae, Endomychidae, Melandryidae and Cryptophagidae (Fig. 342). Several other families have been found on carrion as casual predators or as accidental vistors, but they are unlikely to have any forensic significance.

Another way in which beetles may be involved in forensic investigations is in cases of poisoning. According to Simpson (1980) the vast majority of poisonings are suicidal or accidental and murder by poison is rare.

Cantharides are probably the best known insect poisons and these are obtained from the beetle *Lytta vesicatoria* (L.) (Meloidae) (known as the Spanish-fly) and related beetles. These cantharides are taken in powder form or as a tincture intended as an aphrodisiac, abortifacient, a supposed tuberculosis cure, or by mistake. Recognizable fragments of the beetles have been found in the human stomach months after interment (Gonzales *et al.*, 1954; Keh, 1985).

Arrow poisons may also assume an importance in forensic medicine in certain countries. Many of these are derived from plants, but the bushmen of the Kalahari Desert use an arrow poison derived from pupae of the beetle *Diamphidia nigroornata* Stål (= *locusta* Fairmaire) (Chrysomelidae). Crowson (1981) records other genera as sources of arrow poisons including *Blepharida*, *Cladocera* and *Polyclada* (all Chrysomelidae, Halticinae). Darwin found that a species of Meloidae (unidentified) was used for arrow poison in Cudico, S. Valdivia, Chile (K. G. V. Smith, in press).

Lepidoptera
Butterflies, moths and caterpillars

(Figs 358–365, 370)

This order contains the butterflies and moths and is, in some ways, probably the best known insect group. Lepidoptera have four wings which, along with the body and appendages, are clothed with scales. The adults are mostly nectar-feeders and the larvae mostly plant-feeders.

It has long been known that many butterflies and moths are attracted to carrion, excrement and urine and these substances have been widely used by collectors as baits for these insects. The majority of Lepidoptera visit carrion as adults, to suck up the liquids exuding from the carcase through their proboscis, but there is evidence that some are exclusively carrion feeders.

Reed (1958) recorded the following families attracted to dog carcases in the USA (Tennessee): Phalaenidae (= Noctuidae), Thyrididae, Sphingidae, Geometridae, Hesperiidae, Papilionidae, Nymphalidae and Lycaenidae. He also found a few larvae, but regarded these as accidental occurrences. Payne & King (1969), also working in the USA, found the following families visiting dead pigs: Papilionidae, Satyridae, Nymphalidae, Hesperiidae, Sphingidae, Noctuidae, Geometridae, Pyralidae and Tineidae. All visited the carrion, as adults, in the advanced decay stage, except the Tineidae which bred in dry carcases.

The fungus *Phallus impudicus* Persoon has a strong carrion smell and attracts many carrion-feeding insects as discussed elsewhere. Luther (1946) records *Autographa gamma* (L.) (Noctuidae) visiting the fungus in Finland and Smith (1956) found the noctuid, *Conistra vaccinii* (L.), on the same species of fungus at dusk in November 1954 in England; therefore, these species may visit carcases.

The only two lepidopterous families regularly associated with the faunal succession on cadavers are the Pyralidae and the Tineidae. They may be identified, along with other Lepidoptera associated with stored products, by the detailed keys in Corbet & Tams (1943), Hinton (1956) and Hinton & Corbet (1975), and the Tineidae (clothes-moths) are treated by Austen & Hughes (1948) and other editions and Robinson (1979). Carter (1984) deals with pest species of European Lepidoptera and provides good illustrations, and Busvine (1980) is also useful.

Pyralidae

Aglossa caprealis Hübner (the small tabby-moth, Fig. 358) is a small (2 cm wing span) reddish-brown moth of rather local occurrence, but with a cosmopolitan general distribution. It is normally found in July and August around barns and stables, especially in marshy areas. The eggs hatch in ten days to three weeks and the larvae would normally feed on the damp refuse of wheat stacks, maize and sedge thatch, etc. The larvae live in silken refuse-covered galleries both in and under the material on which they are feeding, usually for nearly two years. Pupation takes place in a tough white silken cocoon which has bits of straw or other refuse attached; the moth emerges in about a month. The larva (Fig. 370) is bronzy-black and slightly greenish tinged. The head and thoracic plate are deep reddish brown.

Aglossa pinguinalis (L.) (the large tabby-moth or grease-moth, Fig. 359) is a similar but larger moth (*c.* 3.5 cm wing span), more variable in size, colour and markings. The species is widely distributed throughout the British Isles and is found in June and July in much the same places as *A. caprealis* but is said to feed on butter and cheese in Europe. Abroad it extends to Europe, North Africa and Cameroun, with other subspecies further east, and has a similar life history.

These species of *Aglossa* are associated with the third wave of Mégnin, when the fats are rancid, some 3–6 months after death (Mégnin, 1894; Leclercq, 1969) and *A. caprealis* may occur in the seventh wave when the corpse is completely dried up.

Other Pyralidae occasionally reared from dried material of animal origin include some species of *Ephestia* and *Plodia interpunctella* (Hübner) (Fig. 360) which is dealt with more fully in the section on Cannabis Insects (p. 169).

Figs 358–365 Lepidoptera (moths) adults. 358, *Aglossa caprealis* (Pyralidae); 359, *A. pinguinalis*; 360, *Plodia interpunctella* (Pyralidae); 361, *Tineola biselliella* (Tineidae); 362, *T. pellionella*; 363, *Monopis rusticella* (Tineidae); 364, *Hofmannophila pseudospretella* (Oecophoridae); 365, *Endrosis sarcitrella* (Oecophoridae).

Tineidae

(Figs 361–362)

Tineola bisselliella Hummel (the common clothes-moth, Fig. 361) is a small (7 mm long) shining, golden-coloured moth. It is a common species of cosmopolitan distribution and may be found at any time of year, but is usually seen in the early summer and autumn in the British Isles. The larvae of this and *T. pellionella* (see below) feed on wool, furs, feathers, dried skins and leather and, in nature, on debris in the nests and burrows of living animals, and on dried animal remains.

The female lays from 40 to 70 eggs over a period of up to 24 days. These hatch in seven days at 24–27°C (75–80°F) and/or up to 37 days at 13°C (55°F). The whitish larvae may feed freely on the surface of their chosen food, but frequently construct protective tubes of silk and food fragments or fibres, inside which they feed from either end. The period spent as a larva varies greatly with temperature and humidity, and may last from two to three months in a favourable environment, but can last up to four years in really harsh conditions. When fully grown the larva makes a strong cocoon of silk and food fibres, inside which it pupates. In 11 to 54 days, dependent on temperature, the adult emerges. This species frequently attacks dried anatomical specimens. Further details of the biology are given by Austen & Hughes (1948).

Tineola pellionella (L.) (the case-bearing clothes-moth, Fig. 362) is a medium-sized (1.00–1.25 cm), dusky-brown moth with three dark spots on the fore wings. It is common throughout the British Isles from June to October, and occurs in Europe and North Africa. Its common name refers to the habit of spinning a portable silken case in which it lives. If separated from this case it may not succeed in building a new one and will die. Like the preceding species, there is considerable variation in the duration of the immature stages and it is not clear if there are one or more broods in the year. The larvae of this species have the reputation of feeding on certain poisonous foodstuffs and surviving, e.g. *Aconitum* root, *Strophanthus* (used for arrow poison in East Africa), laurel leaves, and hemp (*Cannabis*), a habit which may give them an additional forensic significance (see section on *Cannabis* Insects, p. 169). Robinson (1979) provides a revision of the *T. pellionella* complex of species.

Monopis rusticella Clerck (Fig. 363) is a small (6.5–8.5 mm), dark greyish-brown moth with a faint purplish tinge. It occurs through Europe to eastern Asia and the USA. The larvae are detritus feeders, usually on dried animal material in birds' nests, owl pellets, carrion etc. My colleague Dr R. P. Lane has found this species on sheep bones in Wales. Disney (1973) found it on dead shrews, and Hinton (1956) records it in fur of a dead cat, rabbits and crows.

All of these Tineidae are associated with the seventh wave of Mégnin when the corpse has completely dried up. Hinton (1956) provides keys to larvae (see also Carter, 1984).

Oecophoridae

(Figs 364–365)

Other moths which may occur indoors and therefore come to the notice of forensic investigators are *Hofmannophila pseudospretella* (Stainton) (the brown house-moth or false clothes-moth, Fig. 364) and *Endrosis sarcitrella* (L.) (the white-shouldered house moth, Fig. 365). These moths belong to the family Oecophoridae and their larvae feed on debris between the floorboards, in cracks under skirting-boards, etc. They can bite their way through most packing materials. *Hofmannophila* larvae will feed on clothing and soft furnishings but those of *Endrosis* rarely do so.

Outdoors they are found in birds' nests or anywhere where there are accumulations of organic debris. They have not been recorded in the literature as directly associated with human corpses.

Hymenoptera

Bees, wasps, ants, etc.

(Figs 366–369)

The order Hymenoptera includes the familiar bees, wasps and ants, and the less familiar but numerically larger families which live as parasites either of other insects, e.g. Braconidae and Ichneumonidae, or on plants, e.g. Symphyta (saw-flies), where some form galls, e.g. Cynipoidea (gall-wasps).

An introduction to the morphology and taxonomy of the order is given by Richards (1977). Its members may be recognized by the presence of four membranous wings, the fore and hind pairs being hooked together so that they appear as and function as one wing on each side of the body. The abdomen is often 'waisted' (where it joins the thorax) and in several groups the ovipositor is modified to form a sting.

During very intensive studies of pig carcases in the USA (South Carolina), Payne & Mason (1971) recorded an amazing 82 species of Hymenoptera associated with the carrion community. However, few of these have direct forensic significance and only the more familiar Hymenoptera usually in evidence on all carcases are included here.

Wasps (*Vespula* etc.)

These are so familiar that there is no need to describe or illustrate them for the purpose of identification. They appear quite early on in the sequence and will take bits of flesh, especially the eye, from a fresh carcase. However, they play a much more important role as predators and they feed on eggs, larvae and adult flies (including copulating pairs). Since wasps are opportunist feeders upon what is most commonly available, their status in the cadaver succession varies from scavenger to predator.

In addition to their role in the carrion fauna, wasps have been involved in forensic investigations because of their sting. The venom injected by bees and wasps can prove fatal to some sensitive and allergic people and to almost anyone if multiple stinging occurs. Some medical and legal case histories are cited by Leclercq (1969) including one of forensic interest in which 'unnatural parents had shut up an infant in a room full of wasps in order to get rid of it'. Predaceous families of wasps recorded at carrion (other than Vespidae) are Pompilidae and Sphecidae (predators on spiders; see Payne & Mason, 1971).

Bees (Apidae, etc.)

Payne & Mason (1971) found that *Apis mellifera* L. (the honey bee, Fig. 368), *Bombus* and *Xylocopa* were attracted to carrion only while fluids were present, when they were observed to suck up the foul-smelling juices. The ancient legend of bees breeding in carcases is dealt with under *Eristalis* (Diptera, Syrphidae, p. 80).

Like wasps, the sting of bees has led to their forensic implication and Roch (1948) records a case where a nurse put a bee in the mouth of a baby. An unusual case, reported to me by the Metropolitan Police Forensic Science Laboratory, involved the grounding of three aircraft due to the nests of leaf-cutter bees having immobilized the air speed indicators; suspicion of sabotage had been aroused. The shock and pain caused by the stings of bees and wasps may also be responsible for some traffic accidents, especially if an allergic reaction in the driver is involved (Leclercq, 1969).

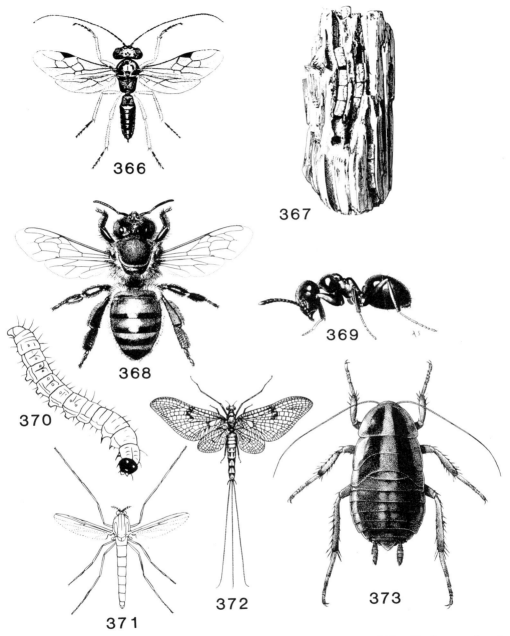

Figs 366–373 Miscellaneous insects of forensic importance. 366, *Alysia manducator* (Hymenoptera, Braconidae); 367, cells of leaf-cutter bee *Megachile willoughbiella* (Megachilidae Hymenoptera); 368, worker honey bee (Apidae, Hymenoptera); 369, ant (Formicidae, Hymenoptera); 370, caterpillar of moth *Aglossa*; 371, chironomid midge (Diptera); 372, mayfly (Ephemeroptera); 373, cockroach, *Blattella germanica* (Dictyoptera).

Ants
(Fig. 369)

Ants can be present at all stages of carrion decomposition. Like wasps they are opportunist feeders on whatever is most readily available. Thus their role in the faunal succession varies from predator (on the eggs and larvae of other insects) to scavenger on the flesh or exudates from the corpse itself. Fuller (1934) felt that the ants were an unimportant element in the carrion fauna since they were present all the time. However, Payne & Mason (1971) held the opposite view and found that ants of several species arrived 3–6 hours after placement and were active carrion feeders, on blood and moist skin, even after the fly maggots had left. Ants are much more important on carrion in tropical than in temperate conditions. Cornaby (1974) found 29 species (*Camponotus* and *Pheidole*) on dead lizards and toads in Costa Rica. Utsumi (1958) found ants dominant on carcases of dogs, rabbits and rats in Japan. On buried pigs in the USA, Payne *et al.* (1968*a*) found that ants (*Prenolepis imparis* Say), along with sphaerocerid and phorid flies, dominated the fauna. Müller (1975) notes that the injuries inflicted on the skin of fresh corpses by the feeding action of ants consist of small gnawed holes which can be misinterpreted as resulting from strong acids. Ants do of course secrete formic acid.

Parasitic Hymenoptera

Alysia manducator (Panzer) (Braconidae; Fig. 366) is the most conspicuous of the parasitic Hymenoptera likely to be seen at carrion in the British Isles. It is a black insect some 6–10 mm long and it lays its eggs on blowfly maggots, where they develop as solitary internal parasites. Eventually the adult emerges from the blowfly puparium. The species parasitizes *Calliphora, Lucilia, Phormia* and species of the non-British genus *Chrysomya*. Graham-Smith (1916) made detailed observations on this insect. The effect of parasitization of blowfly larvae by this species could have forensic significance since it affects the physiological processes involved in pupation by the host. Some substance injected at the time of oviposition induces premature pupation (Holdaway & Smith, 1932; Clausen, 1940). Other effects of parasites on the physiology and resultant development and behaviour of cadaver insects would well repay investigation. The following works should be useful in initiating such work: Clausen (1940), Askew (1971) and Smith (1974). The unpublished work (thesis) of Simmonds (1984) should also be consulted when available.

Other parasitic families of Hymenoptera reported from carrion by Payne & Mason (1971) are the Ichneumonidae, Eulophidae, Pteromalidae, Cynipidae, Figitidae, Evaniidae, Pelecinidae, Proctotrupidae, Diapriidae and Mutillidae. Working with buried flesh in Germany, Lundt (1964) found adult Braconidae and Proctotrupidae in company with phorid flies at a depth of 50 cm.

Parasites have been used in forensic investigations (Leclercq, 1975; Leclercq & Tinant-Dubois, 1973), but the approach is new and further research is needed.

Dictyoptera, Orthoptera, etc.

Cockroaches, crickets, grasshoppers
(Fig. 373)

Cockroaches (Fig. 373) are found in a variety of habitats, including vegetation, and under bark, stones and logs, but are not usually associated with carrion in cool, temperate regions.

Payne (1965) found *Parcoblatta* sp. on pig corpses in South Carolina, USA, during the active decay stage (fourth day), but numbers decreased during the advanced decay stage (sixth day) and when the carcase reached the dry stage (eighth day) they took up positions beneath the carcase. Cornaby (1974, and see Fig. 7), working with toad and lizard carrion in Costa Rica, found *Euphyllodromia angustata* (Latreille), *Lobodromia* sp., *Neoblattella fraterna* (S. & Z.), *Nesomylarcris* sp., and *Nyctobora noctivaga* Rehn during the early stages of decay. Cockroaches were also recorded by Jirón & Cartín (1981) from a dead dog in Costa Rica.

Crickets are usually found in concealed places, in burrows under logs and stones, and in leaf litter and other organic debris. Cornaby (1974) found *Nemobius* sp. and *Niquirana* sp. on toad and lizard carcases in Costa Rica. Reed (1958), in his study of dog carcases in the USA (Tennessee), regarded the presence of 'Orthoptera', including crickets and cockroaches, as accidental.

The role of 'Orthoptera', especially cockroaches and crickets, in carrion decomposition requires further investigation, particularly in concealed indoor situations where these insects are more likely to occur and to which the normal successional fauna may not have access.

Literature useful for identification is rather scattered and is listed by Hollis (1980). Ragge (1965, 1973) treats the British species. Biological information can be found in the general works cited in the introduction.

Hemiptera

Plant-bugs, aphids, etc.

(Figs 374–379)

The majority of Hemiptera are plant-feeding, hence their popular name of plant-bugs. They have piercing suctorial mouthparts, which are usually prominent. Some members of the order use these for killing prey and others for bloodsucking (e.g. the bed-bug (*Cimex*) and the Reduviidae). They have two pairs of wings, but the anterior pair is often of a harder consistency than the posterior pair. The wings are kept folded until required for flight and the insects thus bear a superficial resemblance to beetles.

Coreidae

In general members of this family are plant-feeders, but Payne *et al.* (1968b) found three species feeding on a pig carcase in the USA. *Megalotomus quinquespinosus* (Say) (Fig. 374), *Alydus eurinus* (Say) (Fig. 375) and *A. pilosulus* (Herrich-Schaeffer) were all observed to feed on the carrion. The first two species were present, including mating pairs, as early as the bloated stage of decomposition. Reed (1958) also found the first two species on dog carcases and other American workers have recorded all three species on carrion of various sorts.

Lygaeidae

These are primarily seed-feeders, but some appear to be predaceous. Payne *et al.* (1968b) found *Myodocha serripes* Olivier (Fig. 376) feeding on a calliphorid larva (Diptera) on pig carrion.

Reduviidae

These are predaceous and some feed on the blood of mammals, including man. Bornemissza (1957) recorded *Pirates* sp. hunting around carcases of guinea pigs in Australia. Payne *et al.* (1968a) found *Oncocephalus geniculatus* Stål (Fig. 378), *Melanolestes picipes abdominalis* (Herrich-Schaeffer) (Fig. 379) and *Sinea diadema* (F.) (Fig. 377) on pig carrion during the later stages of decomposition. All three species were observed feeding on adult calliphorids, but only *M. picipes* was seen to feed on dipterous larvae. Payne *et al.* (1968b) also found Cydnidae, Tingidae and Miridae on pig carrion, but they probably had no direct connection with it. Other families (e.g. aphids) may occur accidentally on carrion, especially if a corpse is overhung by vegetation (see Insects and Transport, p. 167).

It is possible that the degree of blood digestion in bed-bugs collected near a corpse could give clues to the post mortem interval, but I know of no published case.

The British species of plant-bugs (suborder Heteroptera) can be identified with the keys of Southwood & Leston (1959), and a key to world genera is provided by Ghauri (*in* Smith, 1973b). Slater & Baranowski (1978) deal with North American species. Aphids should be submitted to a specialist for identification.

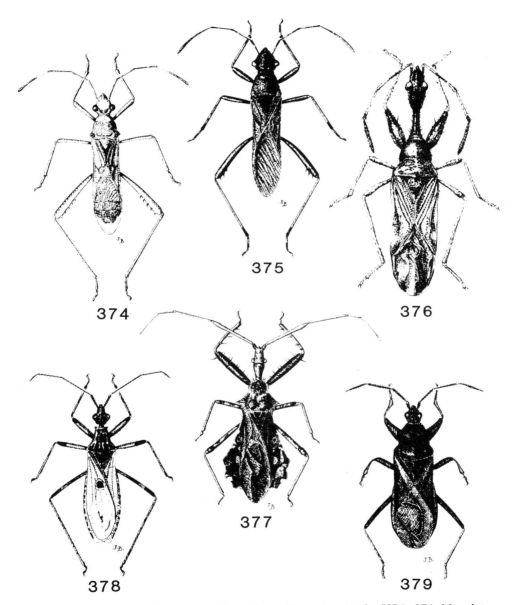

Figs 374–379 Hemiptera-Heteroptera (bugs) on pig carrion in the USA. 374, *Megalotomus quinquespinosus* (Alydidae); 375, *Alydus eurinus* (Alydidae); 376, *Myodocha serripes* (Lygaeidae); 377, *Sinea diadema* (Reduviidae); 378, *Oncocephalus geniculatus* (Reduviidae); 379, *Melanolestes picipes* (Reduviidae). (From Slater & Baranowski, 1978.)

Collembola

Springtails
(Fig. 390)

The Collembola or springtails are so called because of the presence of a forked ventral springing organ on the fourth abdominal segment, which enables them to jump. They are small, wingless insects, rarely exceeding 5 mm in length, and are found in the soil, in decaying vegetable matter, under bark, dead leaves, etc. Some occur in ants' nests or on the surface of water. They vary in coloration from blue-black to green or yellowish.

They are occasionally found on healthy persons in the hair (head and pubic), where they probably feed on the sebum and persist for some time. However, they are more likely to come to forensic notice by their presence on corpses.

Folsom (1902) discusses the Collembola collected from graves in Washington D. C. by Dr M. G. Motter and recorded in his classic paper on the fauna of the grave (Motter, 1898). From over 150 disinterments, 5 600 specimens of Collembola were collected representing six species, five of which were new to science. One species, *Isotoma sepulchralis* Folsom, comprised 97% of the collection. Collembola occurred whenever the environment was in any degree moist, but almost never in dry conditions. They are also found on cadavers exposed on the surface, but they appear to be of little forensic significance.

The three species of Collembola found on a fox corpse, some 45 days after death (Smith, 1975) were herbivorous, although one of them, *Hypogastrura bengtssoni* (Axelson), prefers a more succulent type of rotting vegetable matter. Payne *et al.* (1968a) found Collembola appearing for the first time on pigs in their fourth, 'advanced decay' stage (after six days).

In Australia, Bornemissza (1957) found that subterranean Collembola, such as *Tullbergia* and *Onchiurus*, were intolerant of concentrated decomposition products and soon vacated the soil beneath guinea pig carcasses. However, they were the first of the redeveloping soil fauna to return some eight months later, following heavy rains.

Isoptera

Termites

Termites, or 'white ants' as they are sometimes called, are mostly tropical and subtropical insects, with a few species in warm temperate regions. They are polymorphic social insects with winged sexual forms, and worker and soldier castes. Many species build elaborate nests or mounds and are rarely found away from them.

Termites are primarily cellulose feeders attacking sound or decaying wood, but they will also feed on grass, seeds, leaves, lichens, fungi, topsoil and dung. Their predilection for wood and timber products has brought them into contact with man and earned them a reputation as major pests in his buildings.

M. Coe (1978) found *Odontotermes zambesiensis* Sjöstedt on elephant carcases in Kenya. The termites constructed soil foraging channels over the surface of the bones and were primarily responsible for the removal of the remains of ligaments and cartilaginous caps on epiphyses, after all the skin had disappeared from the corpses.

In Panama, Thorne & Kimsey (1983) found *Nasutitermes nigriceps* (Haldeman) on the dismembered bones and skin of three-toed sloth carcases and a decomposing turtle shell. In all cases the termites had constructed foraging galleries over portions of the skeletons and, in one sloth carcase, were present in the pelvis and cranium and in the marrow cavities. The same authors exposed an agouti carcase on 29 March 1981 and found that the following scavengers were attracted on 31 March: calliphorid, richardiid, muscid and micropezid flies, meloponine bees and scarab beetles. By 13 April *Nasutitermes* had established a foraging gallery. By this time most of the agouti's internal organs and musculature had gone, but the hide was intact. The termites entered the body cavity through a skin tear in the foot and congregated in the cranium. The authors note that termite exploitation of carrion may require dry conditions for scouting and access to a body and may also reflect their nutritional needs in the dry season.

A useful work on termites with a key to families is that by Harris (1971) and regional taxonomic works are listed by Hollis (1980).

Ectoparasites

Siphonaptera (fleas) and Phthiraptera (lice)

(Figs 381–383)

Siphonaptera (fleas, fig. 383) are small, wingless insects, which are laterally compressed and live as ectoparasites on warm-blooded animals. In Britain the human flea (*Pulex irritans* L.), the cat flea (*Ctenocephalides felis* Bouché) and the dog flea (*C. canis* Curtis) are the species most likely to be found on man.

Human fleas are now much less common than previously and their decrease in recent years may be due to improved housekeeping and the use of vacuum cleaners. They also feed on the domestic pig and may infest pigsties (Pig flea would in fact be a better name since primates do not have specific fleas).

Fleas would leave the body of the host shortly after death except in cold environmental conditions and are only likely to be of forensic importance if found on a drowned corpse. Simpson (1985) states that fleas are drowned in 25 hours or so; if immersed for 12 hours they require about an hour for revival and after 18 to 20 hours immersion, a period of some four to five hours is needed for revival.

Eggs are laid in the host animal's nest or sleeping quarters or in its fur, feathers or clothing. The larvae live in and around the nest and living quarters. They are not parasitic, but feed on miscellaneous organic matter and can also develop in debris accumulated in the grooves between floorboards. The larvae are fully developed within one month, when they spin cocoons and change into pupae from which adult fleas emerge, under suitable conditions, within two weeks. Adult fleas may stay inactive, in their cocoons, for as long as a year, until stimulated by vibrations or the movements or pressure of a passing animal. This behaviour explains why the first person to enter a building which has been uninhabited for a long time may be attacked by innumerable fleas.

F. Smit (1957) should be consulted for the identification of the British species, Smit (*in* Smith, 1973b) for world species, and Busvine (1980) gives further biological information.

Phthiraptera (lice, Figs 381, 382) are small, flat, wingless insects which live as ectoparasites on birds and mammals. Man can be infested by three types of lice. The head- and body-lice (Fig. 381) are regarded as varieties of a single species *Pediculus humanus* var. *capitis* De Geer and var. *corporis* De Geer, and the pubic-louse (Fig. 382) belongs to a different genus (*Phthirus pubis* L.). The head-louse may occur on the body, including the pubic region, but the body-louse is never found on the head. The pubic-louse may be found on other parts of the body including the head. The head louse is usually smaller in size, but tougher in texture and darker in colour with a narrower thorax and the abdominal clefts more deeply incised. The pubic-louse (Fig. 382), or 'crab', is smaller than *P. humanus*, but the body is much broader and it is much rarer.

Lice are not restricted to any particular zoogeographic region nor are any races of man more prone or more immune to infestation. Lice may be found anywhere man makes his abode and neglects his standards of hygiene to a degree favourable to their survival. They

Figs 380–391 Insects and other Arthropods of forensic importance. 380, Trichoptera (caddis-flies), larva in case (courtesy of N. Hickin); 381, *Pediculus humanus* (body-louse, Phthiraptera); 382, *Phthirus pubis* (crab-louse); 383, *Pulex irritans* (human flea, Siphonaptera); 384, spider; 385, *Acarus siro*, nymph (Acari, mites); 386, *Acarus siro*, female; 387, *Tyroglyphus* sp., (Acari, mites); 388, millipede; 389, *Forficula auricularia* (Dermaptera, earwigs); 390, Collembola (springtails); 391, *Gammarus pulex* (Crustacea, freshwater shrimp).

ECTOPARASITES 163

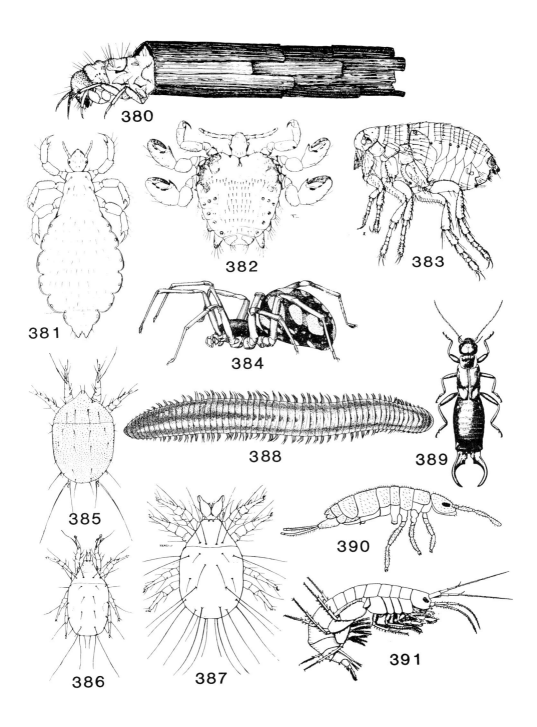

have an incomplete metamorphosis (see Introduction) and both young and adults suck blood. Lice are very sensitive to temperature and will leave the body of a person with a fever as readily as that of a dead person which is becoming cold, though they may stay in clothing if warm conditions prevail (Brisard, 1939). Gonzales *et al.* (1954) state that lice (presumably body-lice) will remain alive on a corpse for three to six days and that if all the lice are dead the person must have died more than six days previous to the examination. They may be of forensic importance if found on a recently drowned body as they usually die about 12 hours after immersion (see Corpses Immersed in Water, p. 25).

Further information on lice may be found in Busvine (1980). See also Arachnida (p. 165) for ectoparasitic mites.

Arthropods other than insects

Millipedes, spiders, mites, etc.

Diplopoda

(Millipedes, Fig. 388)

Millipedes (Diplopoda) are soil-dwelling Arthropoda with two pairs of legs on most of their many abdominal segments (diplo-segments). There are some 8 000 species of which 44 occur in Britain. Millipedes are normally regarded as feeding on decaying plant tissue, fungi and animal excrement.

Hoffman & Payne (1969) have reviewed the carnivorous habits of millipedes following work on pig carcases in the USA. Payne & Crossley (1966) found ten species of millipedes representing seven orders on or under the bodies of baby pigs in South Carolina. These mostly occurred in association with the ultimate stages, 'dry' and 'remains', and were probably in the main only taking cover. However, they did find one species, *Cambala annulata* (Say), regularly on corpses in a state of 'advanced decay' and this species has been recorded in carrion by others.

In studies on buried pigs Payne et al. (1968a) also found *C. annulata* on a corpse buried during the winter months (September 1966 to April 1967). The pig was in an advanced state of decay with the tissues moist and gluey and the millipedes were inside the intact skull, hollow chest cavity and within pockets in the abdomen. The common millipede *Blaniulus guttulatus* (F.) is also known to feed on a wide range of vegetable and animal matter.

Keys to the British species are provided by Blower (1958).

Arachnida

Spiders, mites, etc.

(Figs 384–387)

Spiders (Araneae, Fig. 384) may be found seeking shelter on carrion, especially among the skeletal remains, where they may adopt a predatory role and affect competition. Payne (1965) found 34 species associated with pig carrion in the USA, though these were seldom present in numbers and then only during the active and advanced stages of decay, when insect activity was at a maximum.

Mites (Acari) may occur on carrion in the later, dry, stages of the faunal succession and are usually associated with human corpses 6–12 months after death (Mégnin's sixth wave, see Table 1, Figs 3, 5–7). These mites are usually species associated with stored products, e.g. *Acarus siro* L. (Figs 385, 386), *Tyroglyphus* spp. (Fig. 387), *Glycyphagus* spp. (see also section on *Cannabis* Insects, p. 169).

Immature mites have six legs and adults have eight. They are flightless and many are transported to carrion on the bodies of beetles and flies. Some species are predaceous. Springett (1968), during his studies on competition between blowflies (*Calliphora*) and burying beetles (*Nicrophorus*) on carrion, found that mites were important predators of blowfly eggs. He showed that when there were no mites the beetles were unable to produce any offspring if in competition with more than 100 eggs or larvae of *Calliphora*. If the beetles carried more than 30 mites between them all the blowfly eggs were destroyed and both beetles and mites thrived. Once the blowfly larvae hatched, however, the mites were unable to kill them and both mites and beetles were prevented from breeding.

Mites may also be of importance in indicating where a body, or a suspect, or suspect vehicle (see Insects and Transport) has been. Mites may be present on the vegetation or in the soil on a site of forensic interest and thus provide links with that site if present in samples. Webb *et al.* (1983) report a case in California where the presence of chigger bites (*Eutrombicula belkini* Goula) on a suspect linked him to the scene of a murder where the investigating team were being similarly bitten.

The hair follicle mites, *Demodex*, are host specific to man. Norn (1971) found, in Copenhagen, that 89% of the cadavers of older people harboured *Demodex* in their eyelashes. She further commented that '*Demodex* is tenacious of life only if it is kept in a moist environment. It can survive one week in pure water and two weeks in immersion oil'. Thus *Demodex* may be of use in forensic investigations on bodies submerged for a short period of time. It may also be worth noting that the itch mite *Sarcoptes scabei* L. usually lives for only a few days without feeding and under ideal conditions none will survive for as long as two weeks (Mellanby, 1943).

Mites are very difficult to identify and should be submitted to a specialist. Some information may be gleaned from G. O. Evans *et al.* (1961), Hughes (1961), and Sheals (*in* Smith, 1973*b*). Some forensic cases involving mites have been reported by Leclercq & Watrin (1973). The mite section in Mégnin (1894) should also be studied though the nomenclature is now out of date.

Spiders and scorpions are mostly venomous and a few species are dangerous to man. In certain parts of the world the numbers of known fatalities from scorpions may exceed those from the bites of poisonous snakes (e.g. Mexico, North Africa). Children are more seriously affected than adults. Wherever the few really venomous species occur there is the possibility of forensic involvement in cases of sudden death. A survey of the Arachnida of medical importance is provided by Sheals (*in* Smith, 1973*b*).

Crustacea, etc.

Crustacea may feed on corpses in aquatic situations, e.g. *Gammerus* (Fig. 391), *Asellus*, etc. in freshwater situations, or prawns, crabs and cirripedes in marine conditions (see Corpses Immersed in Water, p. 25).

Taxonomic works on invertebrates other than insects are listed under zoogeographic regions in Sims (1980).

Insects and transport

(Figs 367, 371, 372)

Occasionally entomologists will be called upon to identify insects or fragments of insects (or other invertebrates, e.g. mites, spiders) collected from vehicles of transport involved in forensic investigations. Splashes or stains on cars, etc., may need to be examined to eliminate the possibility of their being human bloodstains, though this is readily established by other means. The usual type of enquiry involves the identification of insects found on a car or other vehicle in order to ascertain the type of terrain through which the vehicle may have passed. All sorts of conditions will influence the *number* of insects found on a vehicle, e.g. the relative abundance of insect life which is usually dependent on locality, the season of the year, the meteorological conditions, the speed and design of car and so on. The type of insect one may expect to find on the front of a vehicle is usually a small, flying, soft-bodied insect which does not bounce off, but bursts on impact and thus remains stuck to the vehicle. Many insects are also attracted to light and their remains should be sought on the car's headlamps.

More robust insects may be run over and jammed in the tyre treads or encased in mud on the underside of the car, in the mudguards or even carried on footwear onto the carpets inside the vehicle (or elsewhere). All of these can be of value in reconstructing a possible journey or establishing whether or not a car (or person) has been in a particular area. This can in turn establish whether a vehicle is likely to have been in a locality where a body has been found or may help to locate a missing body. The presence of maggots or puparia in a car may also establish its use in transporting a body from one site to another (Aruzhonov, 1963).

In the USA the 'lovebug' phenomenon is well known. This refers to the large concentrations of the fly *Plecia nearctica* Hardy (Bibionidae), often found as copulating pairs, along highways during May and September. The flies are smashed on windscreens, obscuring the vision of motorists and may cause cars to overheat when radiators become clogged. These flies have been reported at altitudes of 1500 feet (461 m) by aircraft pilots and several miles out to sea by fishermen. Callahan & Denmark (1973a, b) have shown that the 'lovebugs' are attracted to highways by automobile fumes irradiated by ultraviolet light from the sun. However, these flies do occur in natural swarms at these times of year. A similar phenomenon also occurs in the British Isles where species of *Bibio* and *Dilophus* are involved. At least one forensic case is known where *Plecia nearctica* from a car were used as evidence in establishing the location of a suspect in a homicide case in Florida (F. C. Thompson, pers. comm.). Similar cases have occurred involving butterflies and other insects.

Insects may also be tranported alive inside vehicles, aircraft (Whitfield, 1939; E. B. Knipling, 1958; Sullivan, *et al.* 1958) or ships. Cases of myiasis have been recorded on offshore ships (Downes, 1951) and it is possible that corpses transported over long distances and 'dumped' may yield maggots and other insects alien to that particular geographical region.

Evidence of forensic value may also be gathered from stationary vehicles. As mentioned above, many insects are attracted to light and may assemble on car headlights even if the vehicle is stationary, although of course most would probably fly off when the vehicle moved unless trapped in dew or rain (see below).

Apart from insects staining vehicles due to their bodies bursting on impact they can also affect the paintwork of stationary vehicles by egg-laying or excreting on the paintwork. Theron (1972) found that chironomid midges mistook the shining roofs of dark coloured cars for the water surface on which they normally lay their eggs. Dull, black- or light-coloured cars or vehicles with vinyl tops were virtually free of midges.

When attracted the females became trapped in the dew on the cars as they alighted to lay their eggs. Their bodies caused the paintwork to crack and lift, exposing the bare metal beneath. A similar case was recorded in the British Isles (Dear, 1980) but this time a mayfly, *Caenis moesta* Bengtsson (Ephemeroptera), was responsible. These insects also have aquatic immature stages. Females were attracted at night by ultraviolet light around the car park, mistook the shiny surfaces of the cars for water and were trapped during oviposition. Their bodies gradually disintegrated and the eggs remained and became baked hard on the surface and reacted with the paint, causing small circular areas (0.5–1.0 cm in diameter) on which the paint became pitted, bubbled and corroded. These were new cars parked near a very large reservoir. Some 50–60 vehicles were damaged, but some cars had been supplied with a matt grease coating and were not affected.

Vehicles parked under trees may become marked by excrement, e.g. the sweet sticky honeydew secreted by aphids. This in turn may trap insects attracted to its sweetness for feeding purposes.

All of these phenomena may have forensic significance in the right circumstances and may be applicable to aircraft and ships or boats as well as cars.

Another way in which insects and vehicles may be connected in medico-legal investigations is in the case of traffic accidents caused by insects. Stinging insects may cause an accident by attacking a driver while the vehicle is in motion. Allergic reactions following stings may also cause complications for drivers.

Finally, insects may actually immobilize vehicles of transport as happened when jet aircraft were grounded because small bees had blocked the air-speed indicators with their nests (see Hymenoptera, p. 154).

Cannabis insects

Insects associated with *Cannabis sativa* L.

Insects are sometimes found in confiscated *Cannabis* (marijuana) and may be brought to the forensic scientist for examination and an opinion on the possible country of origin based on the identifications. The insects involved are frequently species widely distributed by commerce as pests of stored products, so that, while identification may thus be relatively easy, it nonetheless may not be possible to identify the precise country of origin. Arnaud (1974) gives details of insects and a mite associated with a case of stored *Cannabis sativa* in San Francisco as follows (numbers of Coleoptera not specified):

Diptera
Desmometopa sp. [*varipalpis*] (Milichiidae) (145)
Coboldia fuscipes sp. (Scatopsidae) (11)
Bradysia sp. (Sciaridae) (20)
Drosophila busckii (Drosophilidae) (1)

Lepidoptera
Plodia interpunctella (Pyralidae) (8)

Coleoptera
Ahasverus advena (Silvanidae)
Oryzaephilus surinamensis (Silvanidae)
Cryptolestes ferrugineus (Cucujidae)
C. pusillus (Cucujidae)
Microgramme arga (Lathridiidae)
Typhaea stercorea (Mycetophagidae)

Acari
Macrocheles muscadomesticae (Macrochelidae)

These and other insects recorded from *Cannabis* are discussed below.

Diptera

Scatopsidae (Fig. 392): *Coboldia fuscipes* Meigen is a small (1.5 mm), black fly with short antennae, and the wings with strong anterior longitudinal veins and weak posterior veins. It has a cosmopolitan distribution and a close association with human activities. The larvae, which live in decaying plant and animal material, have the posterior spiracle on the end of two dorsal horns (Fig. 72). The Palaearctic Scatopsidae are monographed by Cook (1969–1974) and other species may well be found in stored *Cannabis*.
Sciaridae: *Bradysia* spp. (Fig. 394) are small (2–4 mm) slender black flies with long antennae and a characteristic, remarkably uniform venation in which the third longitudinal vein forms an evenly branched fork distally. The species breed in a variety of decaying organic matter, in dead leaves or moss and fungi. The larvae are white with a black head capsule (Fig. 71). The British species of Sciaridae are keyed by Freeman (1983) and other species may well be found in stored *Cannabis*.
Stratiomyidae: *Hermetia illucens* (L.) (Figs 73, 393) is a large (15–20 mm) black species with two translucent spots on the abdomen and wing venation with a very small discal

170 A MANUAL OF FORENSIC ENTOMOLOGY

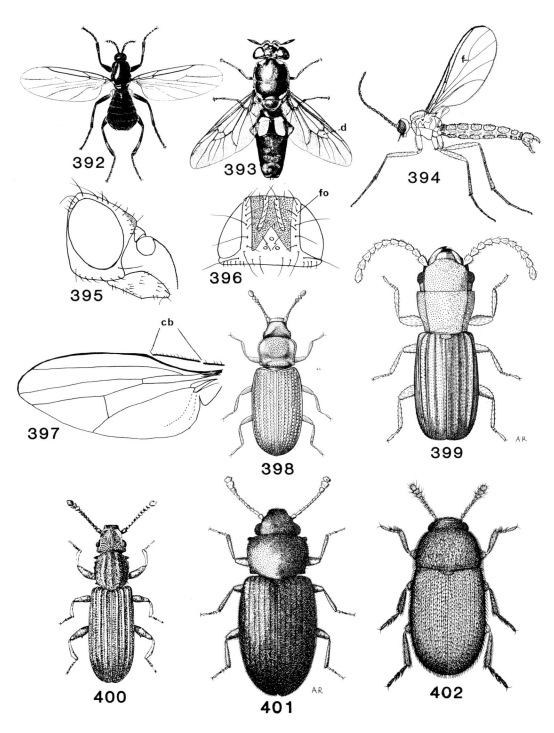

cell characteristic of the family. It has been reared from damp rotting *Cannabis* confiscated by H.M. Customs officials. This fly breeds in a variety of substances, such as decaying fruits and vegetable matter, outdoor lavatories and carrion, including human cadavers. Occasionally larvae are ingested and become involved in intestinal myiasis. A beneficial feature of its larvae is that of rendering faeces in lavatories too liquid for the optimal development of house flies (*Musca domestica*). Although primarily an American species this fly has been transported via commercial trade to Europe, Africa, Asia, the Australian region and some Pacific islands (see p. 77).

Phoridae: *Megaselia* spp. (Figs 78, 85, 294). A very large genus of small brownish flies with a very characteristic venation. Several species have a cosmopolitan distribution and are associated with rotting organic matter generally. One infestation has been seen in association with *Cannabis* in a security area in London.

Milichiidae: *Desmometopa* spp. (Figs 395–398, see also Figs 177–182). These are small (1.0–2.5 mm) black flies distinguished at once by the black, M-shaped marking on the frons (Fig. 396) (N.B. the generic taxonomy of the Acalyptrate Diptera is not normally this straightforward and the 'M' does not stand for marijuana!) Adults are normally found visiting flowers. The larvae have been reared from a wide variety of decaying plant material, manure, dung and sewage. *D. varipalpis* Malloch has been found breeding in sewage filters, septic tanks and faeces of wild silkworm (*Attacus* sp.). Adults are attracted to odours and have occurred in hospital laboratories and operating rooms, dairy cheese rooms, urinals, latrines and butchers, and Arnaud (1974) recorded a huge number 'collected dead in plastic about trunk containing *Cannabis sativa*.' The species is found in all zoogeographic regions as well as on-board aircraft and ships. See Sabrosky (1983).

Drosophilidae (Figs 160–176, 300). These are small flies (2–4 mm) of yellowish to brownish coloration with the antennal arista plumose and ending in a characteristic fork (Fig. 300; see also treatment of the family in taxonomic section, p. 96). They breed in a large variety of rotting and fermenting substances. There are many species with cosmopolitan distribution, including *Drosophila busckii* which has been reared from stored *Cannabis* and has a characteristic larva and puparium (Fig. 161) with a tuberculated body. It is probable that other species will breed in rotting *Cannabis*. The immature stages of the family are described by Okada (1968) and adults occurring in Britain may be identified from the works of Assis Fonseca (1965); Shorrocks' (1972) book is also useful.

There are also Diptera found naturally on living plants of *Cannabis* which may emerge after the plant has been 'harvested' and thus may come to the notice of forensic scientists, e.g. *Phytomyza horticola* Goureau (Agromyzidae), a widely polyphagous leaf-miner, is recorded from Cannabaceae, and Cecidomyiidae (gall midges) may also occur.

Lepidoptera

Pyralidae: *Plodia interpunctella* (Hübner) (the Indian meal-moth; Fig. 360) is a small (6.0–8.5 mm) reddish-brown moth. It is a pest of cereals, dried fruit, seed and nuts, and has a cosmopolitan distribution.

Figs 392–402 Some insects found in stored *Cannabis*. 392, Scatopsidae (Diptera); 393, *Hermetia illucens* (Stratiomyidae, Diptera), d = discal cell; 394, *Bradysia* sp. (Sciaridae, Diptera) (from Freeman, 1983); 395, *Desmometopa varipalpis* (Milichiidae, Diptera), lateral view of head; 396, *D. varipalpis* dorsal view of head (fo = fronto-orbitals, see Identification key to adult flies visiting carrion, p. 68); 397, *Desmometopa m-nigrum*, wing (cb = costal breaks, see Identification key to adult flies visiting carrion, p. 68); 398, *Dienerella filiformis* (Lathridiidae, Coleoptera); 399, *Cryptolestes ferrugineus* (Cucujidae, Coleoptera); 400, *Oryzaephilus surinamensis* (Silvanidae, Coleoptera; 401, *Ahasverus advena* (Silvanidae, Coleoptera); 402, *Typhaea stercorea* (Mycetophagidae, Coleoptera).

Tineidae: *Tineola pellionella* L. (the case-bearing clothes-moth; Fig. 362) is a medium sized (1.0–1.25 cm) dusky-brown moth with three dark spots on the fore-wings. It is common throughout the British Isles and occurs in Europe and North Africa (see also Lepidoptera, p. 152).

Lepidoptera occurring in stored products may be identified by the works of Hinton & Corbet (1975) and Carter (1984, all stages).

Coleoptera

Silvanidae: *Ahasverus advena* (Waltl) (foreign grain-beetle, Fig. 401) is a small (2–3 mm) beetle with a single large 'tooth' on each apical angle of the prothorax. It has a world-wide distribution. Adults and larvae feed on moulds and refuse, but may also attack stored foodstuffs.

Oryzaephilus surinamensis (L.) (saw-toothed grain-beetle; Fig. 400) is a small (2.5–3.5 mm) beetle with six large 'teeth' on each side of the prothorax. It has a world-wide distribution. Adults and larvae attack broken and damaged grain and many other stored food products, particularly of vegetable origin, especially starchy foods. It is a serious pest in bulk storage in Britain and is able to survive in winter even in unheated buildings. The closely related *O. mercator* F. has similar habits, but has not yet been recorded from *Cannabis*. This species has not yet established itself here as it cannot survive our winters.

Cucujidae: *Cryptolestes* spp. (flat grain-beetles; Fig. 399) are small (3 mm) flat beetles with long antennae lacking a distinct club. They have a world-wide distribution and both adults and larvae normally feed on whole and processed grain.

Lathridiidae: *Dienerella* (= *Microgramme*) spp. (Fig. 398) are tiny beetles (1.2–1.6 mm long). Adults and larvae of this family feed on fungi, particularly moulds, and do not attack stored foods though they are commonly found in warehouses, granaries, etc. They are difficult to identify, but seven of the commonest species are keyed in Hinton & Corbet (1975).

Mycetophagidae: *Typhaea stercorea* (L.) (hairy fungus-beetle; Fig. 402) is a small (2.5–3.0 mm) brown, pubescent beetle with a three-segmented antennal club. Adults and larvae feed on fungi and do not attack stored foods although found among them. They also feed on fungi on carrion.

These stored-products beetles are all included in the keys in Hinton & Corbet (1975). Wood-boring beetles of various families may also occur in packing crates and could indicate the country of origin (at least of the crates).

Acari

Macrochelidae: *Macrocheles muscadomesticae* (Scopoli). This is a predaceous mite found on a number of families of Diptera, including *Scatopse fuscipes* (Meigen) with which it was probably associated in Arnaud's (1974) *Cannabis* record.

A case history

In a recent case, Crosby *et al.* (1985, see also Joyce, 1984) found 60 specimens of insects in two separate seizures of cannabis in New Zealand.

Of these only the rice weevil (*Oryzaephilus surinamensis* (L.)) was known to occur in New Zealand, but eight other species were native only to Asia and yielded sufficient information to indicate the precise geographical area as follows.

Coleoptera
  Bruchidius mendosus (Gyllenhall) (Bruchidae) – distributed throughout South East Asia, but is not known from Indonesia or the southern tip of the Malayan Peninsula.
  Tachys sp. (Carabidae) – an abundant tropical genus normally found along the banks of streams or lakes. The specimen found was not Australian.
  Stenus basicornis Kraatz (Staphylinidae) – distributed throughout South East Asia and is usually found on the banks of streams or lakes.
  Azarelius sculpticollis Fairmaire (Tenebrionidae) – a rare species known only from Sumatra and Borneo. It lives as a 'guest' in the nests of termites.
  Gonocnemis minutus Pic (Tenebrionidae) – described from Laos, is known from Thailand and lives as a 'guest' in the nests of termites.

Hymenoptera
  Parapristina verticellata (Waterston) – a pollinator of the fig, *Ficus microcarpa* L., which is distributed throughout the Indo-Australian Region from India and south China to New Caledonia.
  Tropimeris monodon Bouček (Chalcididae) – known from north-west India to Sumbawa in Indonesia.
  Pheidologeton diversus (Jerdon) (Formicidae) – this ant is restricted in distribution to South-East Asia from India to Indochina, including Singapore and West Indonesia. It is commonest in the Indo-Malayan region, including Thailand.

By plotting the distribution of these insects it was possible, from a study of the degree of overlap, to suggest that the *Cannabis* originated in the Tenasserim region between the Andaman Sea to the west and Thailand in the east. From the known habits of the insects it was surmised that the *Cannabis* was harvested near a stream or lake with fig trees and termites' nests nearby. Following presentation of this evidence one of the suspects involved in the case changed his plea from not guilty to guilty.

Batra (1976) records 35 phytophagous species and 14 predaceous species of insects from *Cannabis sativa* var. *indica* Lamarck from three localities in the Kumaon Himalayas. The same author also cites an unpublished review of the literature (by Mostafa & Messenger) in which 272 species of mites and insects are recorded as attacking or otherwise being associated with hemp throughout the world.

Glossary

Acalyptratae (Diptera) – a section of the Cyclorrhapha Schizophora (q.v.).
Acrostichals – bristles or hairs on the dorsal surface (dorsum) of the thorax lying between the dorsocentral bristles (Diptera; Fig. 60, ac).
Adipocere – a fatty, waxy substance resulting from the decomposition of animal corpses in moist places or under water, but not exposed to air.
Aestivation – dormancy during the summer of dry season (cf. hibernation and diapause) (see Fig. 46).
Afrotropical Region – Africa South of the Sahara (including islands) (Fig. 46).
Alar squama (upper Calypter) – a flap-like appendage at the base of the wing in Diptera (Fig. 52).
Anal cell – a cell bounded by the wing veins 5 and 6 in Diptera (Fig. 52).
Anal veins – wing veins 6 and 7 in the Diptera; vein 7, when present, is usually called the axillary vein (Fig. 52).
Antero-dorsal bristles – on the tibia, in front, above (Fig. 51).
Arista – a bristle-like extension of the third antennal segment in the Diptera Brachycera and Cyclorrhapha which may be bare, pubescent or hairy (Fig. 58).
Aschiza (Diptera) – a subdivision of the suborder Cyclorrhapha.
Asynanthropic – associated with man, but in areas devoid of human settlement.
Autolysis – self-dissolution of tissues following the death of their cells, due to the action of their own enzymes.
Basicosta (Diptera) – the second scale-like process on the costal margin of the wing near the base (Fig. 52).
Basitarsus – the first or basal segment of the tarsus, called the metatarsus (Fig. 51).
Biocenosis (pl. biocenoses) – a community of organisms inhabiting a biotope or microhabitat.
Brachycera – the middle of the three suborders of Diptera, linking the primitive Nematocera to the higher Diptera or Cyclorrhapha.
Calypters – see Squamae.
Calyptratae (Diptera) – a section of the Cyclorrhapha Schizophora.
Caterpillar – larva of Lepidoptera (butterflies and moths), usually with three pairs of jointed (true) legs on the thorax and short unjointed prolegs on the abdomen (Fig. 370).
Cell – in wing venation is an area of the wing bounded by veins (Fig. 52).
Chitinized – consisting partly of chitin, the most familiar constituent of the insect endocuticle which renders it tough, but flexible (cf. sclerotized).
Coprophagous – feeding on dung.
Cosmopolitan – with a world-wide distribution, except perhaps for the polar regions.
Costa (Costal vein) – the strong nervure forming the anterior edge of the wing (Fig. 52).
Costal breaks – interruptions in the continuity of the costal vein, an important character in differentiating Acalypterate Diptera (Figs 300, 397).
Coxa – the basal segment of the leg, attached to the thorax (Fig. 49).
Cross-veins – transverse nervures between the longitudinal veins (Figs 52–55).
Cuticle – the horny outer covering of the insect body.
Cyclorrhapha – the third suborder of Diptera which contains the house-flies, bluebottles, greenbottles and their relations ('typical flies').
Diapause – a condition of suspended animation in which development is arrested, usually at a particular stage, to see the insect through unfavourable conditions.
Diel – (rhythms) within a 24-hour period.
Discal bristles, hairs – bristles occurring on the disc or central area of a structure, e.g. abdominal tergites, scutellum or squamae of Diptera.
Discal cell – the cell bounded by veins 4 and 5 in the Diptera (see Fig. 393, d).

Diurnal – active in the daytime.
Dorsocentral bristles (Diptera) – a series of bristles on the dorsal surface (dorsum) of the thorax lying outside the acrostichals (q.v.; Fig. 60, dc).
Dorsum (notum) – the dorsal surface; particularly used in connection with the thorax.
Ecology – the study of the relationships of animals and plants with each other, in communities, and with the physical environment.
Ectoparasite – a parasite which does not completely invade the body, but which feeds superficially on the skin, hair or feathers, or sucks blood.
Endoparasite – a parasite which completely invades the body, i.e. the dermal or subdermal tissues, head cavities or inner organs.
Ethiopian Region – old term for Afrotropical Region (see Fig. 46).
Eusynanthropic – associated with dense human habitations, e.g. cities, towns, etc.
Face – the area of the head surface bounded by the antennae, compound eyes and the anterior edge of the mouth opening.
Facultative parasite – an insect or other animal that is normally free-living, but which, under certain circumstances, may lead a parasitic mode of life.
Family – a grouping of subfamilies and/or genera to form a subdivision of an order or suborder.
Female – is frequently denoted by the sign ♀.
Forensic medicine – medical jurisprudence or the application of medical knowledge to the elucidation of questions in a court of justice.
Frons – the area of the head-surface bounded by the vertex, the compound eyes and the antennae (Fig. 50); sometimes called the front.
Frontal stripe (or frontal vitta) – the area of the frons between the orbits in Diptera Cyclorrhapha.
Fronto-orbital bristles – bristles on the orbits of flies which may be upper (ufo) or lower (lfo) in position (Fig. 50).
Genus (pl. genera) – a grouping of species to form a subdivision of a family or subfamily.
Grub – colloquial name for a larva, usually of the legless maggot type.
Habitat – the locality or situation in which an insect normally lives or in which development takes place.
Haltere (Diptera) – a balancing organ, usually knobbed in appearance and in the position where the hind wings would normally be in four-winged insects (Fig. 49).
Heliotropism – reaction to the sun; may be positively heliotropic (sun loving) or negatively heliotropic (see also phototropism).
Hemisynanthropic – associated with isolated human dwellings.
Hibernation – dormancy during winter (cf. diapause and aestivation).
Holarctic Region – the Palaearctic and Nearctic Regions combined (see Fig. 46).
Humerus (pl. humeri) (Diptera) – one of the anterior corners of the dorsum; a 'shoulder'.
Hyaline (wings) – clear, transparent, translucent, not clouded or coloured.
Hypopleural bristles (Diptera) – those on the hypopleuron, a plate above the coxa of the hind leg (Fig. 49).
Imago (pl. imagines) – the adult sexually developed insect.
Integument – the cuticle or outer covering of the insect body.
Interspecific (e.g. competition) – between species.
Intraspecific (e.g. competition) – within a species.
Larva (pl. larvae) – the active, feeding immature stage emerging from the egg and differing fundamentally in form from the adult.
Larviparous – reproducing by bringing forth larvae which have already hatched inside the female reproductive system.
Lateral bristles – those on the side of a structure; applied to abdominal laterals in Diptera.
Lower squama (or calypter) – the thoracic squama.
Maggot – applied to the typical form of cyclorrhaphous Diptera larva (e.g. Fig. 200).

Male – frequently denoted by the sign ♂.
Median – along the mid-line.
Mesonotum – the dorsum of the thorax.
Metatarsus – the first or basal segment of the tarsus, more correctly the basitarsus.
Micro-climate – the climate in a very small area or habitat.
Myiasis – infestation of live humans and other vertebrate animals with fly (Diptera) larvae, which, at least for a time, feed on the host's dead or living tissue, liquid body substances or ingested food. Subclassified as: *aural* – in the outer ear; *creeping* – burrowing in the skin; *dermal* and *subdermal* – in the dermal tissues or living in boils; *enteric* or *intestinal* – in the alimentary tract; *nasopharyngeal* – in the nasal fossae and the frontal or pharyngeal cavities; *ocular* or *ophthalmic* – in the orbit or eyeball; *rectal* – in the anus and terminal part of the rectum; *sanguinivorous* – living as ectoparasitic bloodsuckers; *traumatic* (or *wound*) – a condition in which the dipterous larvae live as obligate or facultative parasites in traumatic lesions; *urogenital* – in the bladder, the urinary passages or the genital organs. See also pseudomyiasis.
Necrophagous – feeding on dead bodies.
Nematocera – the first of the three suborders of Diptera, containing the mostly gnat-like forms (midges, mosquitoes, crane-flies etc.).
Notum – see dorsum.
Nymph – the immature feeding stage of insects with an incomplete metamorphosis and resembling a miniature version of the adult (cf. larva).
Obligate parasite – a parasite which is dependent on its host and which cannot complete its development without its host.
Ocellar triangle – an area of the frons more or less differentiated from the remainder and on which the ocelli or simple eyes and the ocellar bristles are situated (Fig. 50).
Ocelli – simple eyes, usually arranged in a triangle on top of the head (Fig. 50).
Orbit – the eye margin, often a differentiated strip between the eye and the frons, upon which are situated the orbital bristles (Fig. 50, lfo).
Oviparous – producing eggs.
Ovipositor – the egg-laying apparatus or genital tube of the female through which oviposition is carried out.
Pabulum – food of any kind.
Palp (pl. palpi) – sensory appendages of the proboscis.
Parasite – an organism living on (*ectoparasite*) or in (*endoparasite*) another organism from which it obtains food.
Phototropism – reaction to light, may be positively or negatively phototropic or phototactic (see also heliotropism).
Phytophagous – plant-eating.
Pile – soft down, or short, fine, dense hairs.
Pleuron (pl. pleura) – a lateral sclerite of a segment, usually applied to the thoracic segments; collectively called pleura, but individual plates are denoted by different prefixes, e.g. *hypopleuron*, *sternopleuron*, etc. (see Fig. 49).
Pollinose – covered with 'dust' or 'bloom'.
Post-feeding larva – third stage larva which has irreversibly stopped feeding.
Postvertical bristles – on occiput below the vertex, behind ocellar triangle (Fig. 50, pvt).
Prepupa – the inactive postfeeding larval stage in which the body is contracted and thickened prior to *pupariation* or *pupation* (q.v.).
Prolegs – unjointed abdominal appendages of caterpillars.
Pseudomyiasis – a condition in which living or dead dipterous larvae are found in the intestinal tract of live humans and other vertebrate animals, without having been able to feed or continue their development; such larvae are swallowed with food or drink and do not adopt a parasitic mode of life, but may cause allergic or other reactions as passive 'foreign bodies' in a living animal.

Pseudopod – a false foot, a ventral protuberance or tubercle, characteristic of some dipterous larvae (see Fig. 75).
Ptilinum – an inflatable sac above the base of the antennae in the Diptera Cyclorrhapha Schizophora; used by the emergent fly to rupture the puparium and to penetrate the pupation medium; in the fully developed adult it leaves a ptilinal suture or groove marking the boundaries of the sac.
Pubescence – short fine hair.
Pupa (pl. pupae) – a transformational (metamorphic), but usually immobile, stage between the last larval stage and the adult (*imago*); a chrysalis (Lepidoptera).
Pupariation – when the post-feeding blowfly larva or white prepupa forms its white puparium which tans and hardens to form the dark brown puparium after about 24 hours.
Puparium (pl. puparia) – the thickened tanned, hardened, barrel-like last larval skin inside which the pupa of the higher Diptera (Cyclorrhapha) is formed (Fig. 31).
Pupation – the act of forming the pupa.
Saprophagous – feeding on dead and decaying organic matter.
Sarcosaprophagous — feeding on dead flesh.
Schizophora (Diptera) – a subdivision of the suborder Cyclorrhapha, comprising the Calyptrates and the Acalyptrates.
Sclerite – a more or less sclerotized chitinous plate.
Sclerotized – consisting of the substance sclerotin, a 'tanned protein' which forms the outer, hard, horny layer of the cuticle (cf. chitinized).
Scotophilic – shade loving.
Scutellum – the shield-shaped part of the dorsum lying behind the thorax (Figs 49, 60).
Species – a group of potentially interbreeding individuals which are unable to interbreed with individuals of other such groups.
Spiracle – an external opening of the tracheae or breathing tubes (Figs 37, 49).
Squamae (calypters) (Diptera) – flap-like appendage at the base of the wing (*upper*), or attached to the thorax close to the wing base (*lower*).
Sternite – the ventral plate, or sclerite, of a segment.
Style – a thickened terminal portion (usually tapered) of the third antennal segment in some Diptera (Fig. 57).
Sub-costa – the auxillary vein below the costa (Fig. 52, sc).
Synanthropic – associated with man; see also *asynanthropic, eusynanthropic, hemisynanthropic.*
Tarsus – the segmented foot or apical part of the leg (Fig. 51).
Tergite – the dorsal plate, or sclerite, of a segment.
Tessellate – chequered (see Fig. 306, abdomen).
Thoracic squama (lower calypter) (Diptera) – a flap-like appendage attached to the thorax at the base of the wing (Fig. 52).
Tracheae – the breathing tubes which open at the spiracles.
Transverse suture – a suture or groove (complete or incomplete) dividing the prescutum from the scutum on the dorsal surface of the thorax in some Diptera (Figs 49, 59, 60).
Trochanter – the small segment of the leg between the coxa and the femur.
Tubercle – a small prominence or elevation, more or less rounded.
Veins – the nervures of the wings.
Venation – the arrangement of the wing veins.
Vertex – the top of the head.
Vertical bristles – on top of the head (see Fig. 50, vt = *verticals*, pvt = *postverticals*).
Vibrissae – the large bristles arising from the angles at the side of the mouth in many cyclorrhaphous flies (Fig. 300).
Vitta – a stripe, usually applied to the frons between the orbits; frontal vitta.
Zoogeographical Regions – see Fig. 46, p. 54.

Bibliography

All publications included in the *World List of Scientific Periodicals* are treated as serial publications here, but the titles of journals are given in full. However, where certain serial publications are frequently treated or cited as books additional information on publishers and place of publication is added for clarity. Works cited are in the language of the title unless otherwise stated. A selected bibliography of forensic entomology is provided by Meek *et al.* (1983) and a definitive bibliography by Vincent *et al.* (1985), but these do not include taxonomic references.

Abbott, C. E. 1937. The necrophilous habit of Coleoptera. *Bulletin of the Brooklyn Entomological Society* **32**: 202–204.

Akopyan, M. M. 1953. [The fate of corpses of ground squirrels on the Steppe]. *Zoologicheskii Zhurnal* **32**: 1014–1019 [in Russian].

Aldrich, J. M. 1916. *Sarcophaga and allies in North America* [Vol. 1] 302 pp, 16 pls, La Fayette.

Alwar, V. S. & Seshiah, S. (1958. Studies on the life-history and bionomics of *Sarcophaga dux* Thomson, 1868. *Indian Veterinary Journal* **35**: 559–565

Anderson, J. R. 1963. Methods for distinguishing nulliparous from parous flies and for estimating the ages of *Fannia canicularis* and some other cyclorrhaphous Diptera. *Annals of the Entomological Society of America* **57**: 226–236.

Ardö, P. 1953. Likflugan, *Conicera tibialis* Schmitz, i Sverige (Dipt., Phoridae). *Opuscula Entomologica* **18**: 33–36

Arnaud, P. H. 1974. Insects and a mite associated with stored *Cannabis sativa* Linnaeus. *The Pan-Pacific Entomologist* **53**(1): 91–92.

Aruzhonov, A. M. 1963. The use of entomological observations in forensic science. *Sudebno-meditsinskaya ékspertisa* **6**: 51 [in Russian].

Askew, R. R. 1971. *Parasitic Insects.* xvii + 316 pp, 124 figs, Heinemann, London.

Assis Fonseca, E. C. M. 1965. A short key to the British Drosophilidae (Diptera) including a new species of *Amiota*. *Transactions of the Society for British Entomology* **16**: 233–244.

Assis Fonseca, E. C. M. 1968. Diptera Cyclorrhapha Calyptrata Section (b) Muscidae. *Handbooks for the Identification of British Insects* **10**(4b): 1–119.

Aubertin, D. 1933. Revision of the genus *Lucilia* R.-D. (Diptera, Calliphoridae). *Journal of the Linnean Society, Zoology* **38**: 389–436.

Austen, E. E. 1896. Necrophagous Diptera attracted by the odour of flowers. *Annals and Magazine of Natural History* (6) **18**: 237–240.

Austen, E. E. & Hughes, A. W. McKenny 1948. *Clothes moths and house moths, their life-history, habits and control.* British Museum (Natural History), London, Economic Series 14. [Several editions earlier and later have different information and are all worthy of study.]

Backlund, H. O. 1945. Wrack fauna of Sweden and Finland, ecology and chorology. *Opuscula Entomologica Supplement* **5**: 1–236

Bailey, D. L. 1970. Forced air for separating pupae of houseflies from rearing medium. *Journal of Economic Entomology* **63**: 331–333

Balfour-Browne, F. 1958. *British Water Beetles,* Vol. 3. liii + 210 pp, 87 figs, 67 maps, Ray Society, London.

Barnes, J. K. 1984. Biology and immature stages of *Dryomyza anilis* Fallén (Diptera: Dryomyzidae). *Proceedings of the Entomological Society of Washington* **86**: 43–52

Basden, E. B. 1947. Breeding the house fly, *Musca domestica* L., in the laboratory. *Bulletin of Entomological Research* **37**: 381–387.

Basden, E. B. 1954. The distribution and biology of Drosophilidae (Diptera) in Scotland, including a new species of *Drosophila*. *Transactions of the Royal Society of Edinburgh* **62**: 602–654

Batra, S. W. T. 1976. Some insects associated with hemp or marijuana (*Cannabis sativa* L.) in Northern India. *Journal of the Kansas Entomological Society* **49**: 385–388.

Baumgartner, D. L. & Greenberg, B. 1984. The genus *Chrysomya* (Diptera: Calliphoridae) in the New World. *Journal of Medical Entomology* **21**: 105–113

Beaver, R. A. 1969. Anthomyiid and muscid flies bred from snails. *Entomologist's Monthly Magazine* **105**: 25–26.

Beaver, R. A. 1971. Ecological studies on Diptera breeding in dead snails, 1, Biology of the species found in *Cepaea nemoralis* (L.). *Entomologist* **105**: 41–52

Berndt, K.-P. & Groth, U. 1971. *Protophormia terraenovae* (R.-D.) (Diptera: Calliphoridae) as experimental animals in studies on chemosterilants. *Biologisches Zentralblatt* **90**: 217–222 [in German].
Beyer, J. C., Enos, Y. F. & Stajić, M. 1980. Drug identification through analysis of maggots. *Journal of Forensic Sciences* **25**: 411–412
Bianchini, G. 1930. La biologia del cadavere. *Archivic Antropologia Criminale, Psichiatria e Medicina Legale* **50**: 1035–1105.
Bishopp, F. C. 1915. Flies which cause myiasis in man and animals – some aspects of the problem. *Journal of Economic Entomology* **8**: 317–329.
Blair, K. G. 1922. Notes on the life-history of *Rhizophagus parallelocollis* Gyll. *Entomologist's Monthly Magazine* **58**: 80–83.
Blower, J. G. 1958. British Millipedes (Diplopoda). *Synopses of the British Fauna* **11**: 1–74
Bohart, G. E. & Gressitt, J. L. 1951. Filth-inhabiting flies of Guam. *Bulletin of the Bernice P. Bishop Museum* **204**: vii + 1–152, 17 pls.
Borgmeier, T. 1968, 1971. A Catalogue of the Phoridae of the World (Diptera, Phoridae). *Studia Entomologica* (n.s.) **11**: 1–367; Supplement **14**: 177–224
Bornemissza, G. F. 1957. An analysis of arthropod succession in carrion and the effect of its decomposition on the soil fauna. *Australian Journal of Zoology* **5**: 1–12.
Bøving, A. G. & Henriksen, K. L. 1938. The developmental stages of the Danish Hydrophilidae. *Videnskabelige Meddelelser fra Dansk Naturhistorisk* **102**: 27–162
Braack, L. E. O. 1981. Visitation patterns of principal species of the insect-complex at carcasses in the Kruger National Park. *Koedoe* **24**: 33–49.
Brindle, A. 1962. Taxonomic notes on the larvae of British Diptera. 11. Trichoceridae and Anisopodidae. *Entomologist* **95**: 285–288.
Brindle, A. 1965a. Taxonomic notes on the larvae of British Diptera, No. 20 – the Sepsidae. *Entomologist* **78**: 137–140.
Brindle, A. 1965b. Taxonomic notes on the larvae of British Diptera, No. 21 – the Piophilidae. *Entomologist* **78**: 158–160.
Brindle, A. & Smith, K. G. V. 1979. The immature stages of flies. In Stubbs, A. & Chandler, P. (Eds) A Dipterist's Handbook. *The Amateur Entomologist* **15**: 38–64.
Brisard, C. 1939. Pediculus vestimenti. *Annales de médicine légale de criminologie et de police scientifique* **9–10**: 614–615.
Britton, E. B. 1956. Scarabaeoidea (Lucanidae, Trogidae, Geotrupidae, Scarabaeidae). *Handbooks for the Identification of British Insects* **5**(11): 1–29.
Broadhead, E. C. 1980. Larvae of Trichocerid flies found on human corpse. *Entomologist's Monthly Magazine* **116**: 23–24.
Busvine, J. R. 1980. *Insects and Hygiene*, 3rd edn. viii + 568 pp, Chapman & Hall, London.
Callahan, P. S. & Denmark, H. A. 1973a. Attraction of the lovebug, *Plecia nearctica* Hardy (Diptera: Bibionidae), to automobile exhaust fumes. *Florida Entomologist* **56**: 113–119.
Callahan, P. S. & Denmark, H. A. 1973b. The 'lovebug' phenomenon. *Proceedings of the Tall Timbers Conference of Ecological Animal Control* **5**: 93–101.
Campbell, E. & Black, R. J. 1960. The problem of migration of mature fly larvae from refuse containers and its implications on the frequency of refuse collection. *California Vector Views* **7**: 9–15.
Cantrell, B. K. 1981. The immature stages of some Australian Sarcophaginae (Diptera: Sarcophagidae). *Journal of the Australian Entomological Society* **20**: 237–248.
Carter, D. J. 1984. The Pest Lepidoptera of Europe with special reference to the British Isles. *Series Entomologica* **31**: 1–431, 79 figs, 41 pls, 40 maps.
Chapman, R. F. 1982. *The Insects: structure and function*. 919 pp. Hodder & Stoughton, London.
Chapman, R. F. & Sankey, J. H. P. 1955. The larger invertebrate fauna of three rabbit carcasses. *Journal of Animal Ecology* **24**: 395–402.
Chapman, R. K. 1944. An interesting occurrence of *Musca domestica* L. larvae in infant bedding. *The Canadian Entomologist* **76**: 230–232.
Cheong, W. H., Mahadevan, S. & Singh, K. I. 1973. Three species of fly maggots found on a corpse. *South-East Asian Journal of Tropical Medicine and Public Health* **4**: 281.
Chillcott, J. G. 1961. A revision of the Nearctic species of Fanniinae (Diptera: Muscidae). *The Canadian Entomologist, Supplement* **14**: 1–295.
Chinery, M. 1973. *A Field Guide to the Insects of Britain and N. Europe*. 352 pp, 60 pls, Collins, London.
Chu, H. F. & Wang, L.-Y. 1975. Insect carcasses unearthed from the Chinese antique tombs. *Acta Entomologica Sinica* **18**: 333–337 [in Chinese with English summary].
Clark, C. U. 1895. On the food habits of certain dung and carrion beetles. *Journal of the New York Entomological Society* **3**: 61.
Clausen, C. P. 1940. *Entomophagous Insects*. x + 688 pp, McGraw Hill, New York & London.
Cockburn, A., Barraco, R. A., Reyman, T. A. & Peck, W. H. 1975. Autopsy on an Egyptian mummy.

Science, New York **187**: 1155–1160.
Coe, M. 1978. The decomposition of elephant carcases in the Tsavo (East) National Park, Kenya. *Journal of Arid Environments* **1**: 71–86.
Coe, R. L., Mattingly, P. F. & Freeman, P. 1950. Diptera. 2. Nematocera: families Tipulidae to Chironomidae. *Handbooks for the Identification of British Insects* **9**(2): 1–216.
Cogan, B. H. & Smith, K. G. V. 1974. *Instructions for Collectors. No. 4a. Insects.* 169 pp, 37 figs, British Museum (Natural History), London.
Cole, F. R. & Schlinger, E. I. 1969. *The Flies of Western North America.* 693 pp, 360 figs, University of California Press, Berkeley & Los Angeles.
Colless, D. H. & McAlpine, D. K. 1970. Diptera (Flies) in CSIRO. [Corp. auth.] *The Insects of Australia. A Textbook for Students and Research Workers.* pp. 656–740, Melbourne University Press, Carlton, Victoria.
Collin, J. E. 1943. The British species of Helomyzidae (Diptera). *Entomologist's Monthly Magazine* **79**: 234–251.
Colyer, C. N. 1954a. The 'coffin fly', *Conicera tibialis* Schmitz (Dipt., Phoridae). *Journal of the Society for British Entomology* **4**: 203–206.
Colyer, C. N. 1954b. More about the 'coffin fly', *Conicera tibialis* Schmitz (Diptera, Phoridae). *The Entomologist* **87**: 130–132.
Colyer, C. N. 1954c. Further emergences of *Conicera tibialis* Schmitz, the 'coffin fly'. *The Entomologist* **87**: 234.
Colyer, C. N. & Hammond, C. O. 1968. *Flies of the British Isles* 384 pp, 53 pls, 17 figs, F. Warne, London & New York.
Cook, E. F. 1969–1974. A synopsis of the Scatopsidae of the Palaearctic. I–III. *Journal of Natural History* **3**: 393–407; **6**: 625–634; **8**: 61–100.
Corbet, A. S. & Tams, W. H. T. 1943. Keys for the identification of the Lepidoptera infesting stored food products. *Proceedings of the Zoological Society of London* (B) **113**: 55–148.
Cornaby, B. W. 1974. Carrion reduction by animals in contrasting tropical habitats. *Biotropica* **6**: 51–63.
Cousin, G. 1932. Étude expérimentale de la diapause des insectes. *Bulletin Biologique de la France et de la Belgique* **15**: 1–341.
Cragg, J. B. 1955. The natural history of sheep blowflies in Britain. *Annals of Applied Biology* **42**: 197–207.
Cragg, J. B. 1956. The olfactory behaviour of *Lucilia* species (Diptera) under natural conditions. *Annals of Applied Biology* **44**: 467–477.
Cragg, J. B. & Cole, P. 1952. Diapause in *Lucilia sericata* (Mg.) Diptera. *Journal of Experimental Biology* **29**: 600–604.
Cragg, J. B. & Hobart, J. 1955. A study of a field population of the blowflies *Lucilia caesar* (L.) and *L. sericata* (Mg.). *Annals of Applied Biology* **43**: 645–663.
Crosby, T. K., Watt, J. C., Kistemaker, A. C. & Nelson, P. E. 1985. Entomological identification of the origin of imported cannabis. *Journal of the Forensic Science Society.*
Crosskey, R. W. (Ed.) 1980. *Catalogue of the Diptera of the Afrotropical Region.* 1437 pp, British Museum (Natural History), London.
Crowson, R. A. 1956. Coleoptera: Introduction and keys to families. *Handbooks for the Identification of British Insects* **4**(1): 1–59, 118 figs.
Crowson, R. A. 1981. *The Biology of the Coleoptera.* xii + 802 pp, 317 figs, 15 tables, 9 pls, Academic Press, London, New York &c.
Cuthbertson, A. 1933. The habits and life-histories of some Diptera in Southern Rhodesia. *Proceedings and Transactions of the Rhodesia Scientific Association* **32**: 81–111.
Cuthbertson, A. 1935. Biological notes on some Diptera in Southern Rhodesia. *Occasional Papers of the Rhodesian Museum* **4**: 11–28.
Dahl, C. 1973. Notes on the arthropod fauna of Spitsbergen. III. 14. Trichoceridae (Dipt.) of Spitsbergen. *Annales Entomologici Fennici* **39**: 49–59.
Dahl, C. & Alexander, C. P. 1976. A world catalogue of Trichoceridae Kertesz, 1902 (Diptera). *Entomologica Scandinavica* **7**: 7–18.
Dahl, R. 1969. *Immature Stages of Ephydridae (Diptera, Brachycera).* 66 pp, 198 figs. [Privately circulated.]
Davies, W. M. 1934. The sheep blowfly problem in North Wales. *Annals of Applied Biology* **21**: 267–282.
Davis, W. T. 1928. *Lucilia* flies anticipating death. *Bulletin of the Brooklyn Entomological Society* **23**: 118.
Dear, J. P. 1980. Damage to car paintwork by Ephemeroptera. *Entomologist's Monthly Magazine* **116**: 197.
Dear, J. P. 1981. Blowfly recording scheme – an interim report. *Entomologist's Monthly Magazine* **117**: 75–76.
Dear, J. P. 1985. A revision of the New World Chrysomyini (Diptera, Calliphoridae). *Revista brasileira Zoologia* **3**: 109–169.
Dear, J. P. 1986 Calliphoridae (Insecta, Diptera). *Fauna of New Zealand* **8**: 86 pp., D.S.I.R., Wellington.
Deeming, J. C. & Knutson, L. V. 1966. Ecological notes on some Sphaeroceridae reared from snails, and a description of the puparium of *Copromyza (Apterina) pedestris* Meigen (Diptera). *Proceedings of the Entomological Society of Washington* **68**: 108–112.

Delfinado, M. D. & Hardy, D. E. (Eds). 1973–1977. *A Catalogue of the Diptera of the Oriental Region*. 3 vols, x + 618 pp; x + 459 pp; x + 854 pp, University Press of Hawaii, Honolulu.
Demerec, M. (Ed.) 1950. *Biology of Drosophila*. x + 632 pp. John Wiley & Sons, New York.
Denno, R. F. & Cothran, W. R. 1975. Niche relationships of a guild of necrophagous flies. *Annals of the Entomological Society of America* **68**: 741–754.
Deonier, C. C. 1940. Carcass temperatures and their relation to winter blowfly activity in the Southwest. *Journal of Economic Entomology* **33**: 166–170.
de Stefani, T. 1921. Importanza dell'Entomologia applicata nell' Economia sociale. Entomologia legale e die cadaveri. *Allevamenti, Palermo* **2**: 131–133.
Dethier, V. G. 1947. The role of the antennae in the orientation of carrion beetles to odors. *Journal of the New York Entomological Society* **55**: 285–293.
Detinova, T. S. 1962. Age-grouping methods in Diptera of medical importance. *World Health Organization Monograph Series* **47**: 1–216.
Detinova, T. S. 1968. Age structure of insect populations of medical importance. *Annual Review of Entomology* **13**: 427–450.
Disney, R. H. L. 1973. Diptera and Lepidoptera reared from dead shrews in Yorkshire. *The Naturalist* **927**: 136.
Disney, R. H. L. 1983. Scuttle flies, Diptera, Phoridae (except *Megaselia*). *Handbooks for the Identification of British Insects* **10**(6): 1–81. [*Megaselia* in press.]
Dorsey, C. K. 1940. A comparative study of the larvae of six species of *Silpha*. *Annals of the Entomological Society of America* **33**: 120–139.
Downes, J. A. 1951. Two cases of myiasis on board ship, due to the larvae of *Cordylobia anthropophaga*. *Annals of Tropical Medicine & Parasitology* **45**: 169–170
Draber-Mońko, A. 1973a. A review of the Polish Sarcophagidae (Diptera). *Fragmenta Faunistica* **19**: 157–225 [in Polish with German and Russian summaries].
Draber-Mońko, A. 1973b. Einige Bermerkungen über die Entwicklung von *Sarcophaga carnaria* (L.) (Diptera, Sarcophagidae). *Polskie Pismo Entomologiczne* **43**: 301–308.
Duda, O. 1918. Revision der europäischen Arten der Gattung *Limosina* Macquart (Dipteren). *Abhandlungen der (K.K.) Zoologisch-Botanischen Gesellschaft in Wien* **10**: 1–240.
Duffield, J. E. 1937. Notes on some animal communities in Norwegian Lapland. *Journal of Animal Ecology* **6**: 160–168.
Dunn, L. H. 1916. *Hermetia illucens* breeding in a human cadaver. *Entomological News* **27**: 59–61.
Easton, A. M. 1944. *Lathrimaeum atrocephalum* Gyll (Col., Staphylinidae): a medico-legal problem. *Entomologist's Monthly Magazine* **80**: 237.
Easton, A. M. 1966. The Coleoptera of a dead fox (*Vulpes vulpes* (L.)); including two species new to Britain. *Entomologist's Monthly Magazine* **102**: 205–210.
Easton, A. M. & Smith, K. G. V. 1970. The entomology of the cadaver. *Medicine, Science and the Law* **10**: 208–215, 3 pls.
Egglishaw, H. J. 1960. Studies on the family Coelopidae (Diptera). *Transactions of the Royal Entomological Society of London* **112**: 109–140.
Egglishaw, H. J. 1961. The life history of *Thoracochaeta zosterae* (Hal.) (Dipt., Sphaeroceridae). *Entomologist's Monthly Magazine* **96**: 124–128.
Elton, C. S. 1966. *The Pattern of Animal Communities*. 432 pp. 28 figs, 82 pls, Methuen, London; Wiley, New York.
Emden, F. I. van. 1954. Diptera Cyclorrhapha Calyptrata (1) Section (a) Tachinidae and Calliphoridae. *Handbooks for the Identification of British Insects* **10**(4a): 1–133.
Erzinçlioğlu, Y. Z. 1980. On the role of *Trichocera* larvae (Diptera, Trichoceridae) in the decomposition of carrion in winter. *Naturalist* **105**: 133–134.
Erzinçlioğlu, Y. Z. 1983. The application of entomology to forensic medicine. *Medicine, Science and the Law*. **23**: 57–63.
Erzinçlioğlu, Y. Z. 1985. Immature stages of British *Calliphora* and *Cynomya*, with a re-evaluation of the taxonomic characters of larval Calliphoridae (Diptera). *Journal of Natural History* **19**: 69–96
Erzinçlioğlu, Y. Z. & Whitcombe, R. P. 1983. *Chrysomya albiceps* (Wiedemann) (Dipt., Calliphoridae) in dung and causing myiasis in Oman. *Entomologist's Monthly Magazine* **119**: 51–52.
Evans, A. C. 1935. Studies on the influence of the environment on the sheep blow-fly *Lucilia sericata* (Mg.). II. The influence of humidity and temperature on prepupae and pupae. *Parisitology* **27**: 291–298.
Evans, G. O., Sheals, J. G. & Macfarlane, D. 1961. *The Terrestrial Acari of the British Isles. An Introduction to their Morphology, Biology and Classification. I. Introduction and Biology*. 219 pp, British Museum (Natural History), London. [Only Vol. I published.]
Fabre, J. H. 1913. *The Life of the Fly* [trans. A. T. de Mattos]. 477 pp, Hodder & Stoughton, London.
Fabre, J. H. 1918a. *The Wonders of Instinct* [trans. A. T. de Mattos & B. Miall]. 320 pp, 16 pls, Unwin, London.

Fabre, J. H. 1918b. *The Sacred Beetle and Others* [trans. A. T. de Mattos]. 296 pp, Hodder & Stoughton, London

Fabre, J. H. 1919. *The Glow-worm and Other Beetles.* 488 pp, Hodder & Stoughton, London.

Fabre, J. H. 1922. *More Beetles* [trans. A. T. de Mattos]. 322 pp, Hodder & Stoughton, London.

Fan, C.-T. 1957. Notes on third stage larvae of synanthropic flies in Shanghai district. *Acta Entomologica Sinica* **7**: 405–422 [in Chinese, but illustrated and with English summary].

Fan, C.-T. 1965. *Key to the Common Synanthropic Flies of China.* xv + 330 pp, Academy of Science, Peking [in Chinese].

Ferrar, P. 1979. The immature stages of dung-breeding muscid flies in Australia, with notes on the species, and keys to larvae and pupae. *Australian Journal of Zoology Supplement Series* **73**: 1–106.

Ferrar, P. 1980. Cocoon formation by Muscidae (Diptera). *Journal of the Australian Entomological Society* **19**: 171–174.

Fiedler, O. G. H. 1951. Die Nahrung der Myiasis-Fliegen-larven auf dem Wollschaf. *Zeitschrift für Angewewandt Entomologie* **33**: 142–150.

Folsom, J. W. 1902. Collembola of the grave. *Psyche, Cambridge* **9**: 363–367.

Foreman, F. W. & Smith, G. S. Graham – 1917. Investigations on the prevention of nuisances arising from flies and putrefaction. *Journal of Hygiene* **16**: 109–226.

Forster, H. 1914. *Piophila nigriceps* larvae in a human corpse. *Zoologischer Anzeiger* **45**: 47.

Fourman, K. 1936. Kleintierwelt, Kleinklima und Mikroklima in Beziehung zur Kinnzeichnung des forstlichen Standortes und der Bestandesabfallzersetzung auf bodenbiologischer Grundlage. *Mitteilungen Forstwirtschaft Forstwissenschaft* **7**: 596–615.

Fourman, K. 1938. Untersuchungen über die Bedeutung der Bodenfauna bei der biologischen Umwandlung des Bestandesabfalls forstlicher Standorte. *Mitteilungen Forstwirtschaft Forstwissenschaft* **9**: 144–169.

Fraenkel, G. 1935. Observations and experiments on the blow-fly (*Calliphora erythrocephala*) during the first day after emergence. *Proceedings of the Zoological Society of London* **1935**: 893–904.

Fraenkel, G. & Bhaskaran, G. 1973. Pupariation and pupation of cyclorrhaphous flies (Diptera): terminology and interpretation. *Annals of the Entomological Society of America* **66**: 418–422.

Fraser, A. & Smith, W. F. 1963. Diapause in larvae of green blowflies (Diptera: Cyclorrhapha: *Lucilia* spp.). *Proceedings of the Royal Entomological Society of London* (A) **38**: 90–97.

Fredeen, F. J. H. & Taylor, M. E. 1964. Borborids (Diptera: Sphaeroceridae) infesting sewage disposal tanks, with notes on the life cycle, behaviour and control of *Leptocera (Leptocera) caenosa* (Rondani). *The Canadian Entomologist* **96**: 801–808.

Freeman, P. 1983. Sciarid Flies, Diptera, Sciaridae. *Handbooks for the Identification of British Insects* **9**(6): 1–68.

Freidberg, A. 1981. Taxonomy, natural history and immature stages of the bone-skipper *Centrophlebomyia furcata* (Fabricius) (Diptera: Piophilidae, Thyreophorina). *Entomologica Scandinavica*, **12**: 320–326.

Freude, H., Harde, K. W. & Lohse, G. A. 1965–1964 *Die Käfer Mitteleuropas.* 11 vols, Goecke & Evers, Krefeld.

Fuller, M. E. 1932. The larvae of the Australian sheep blowflies. *Proceedings of the Linnean Society of New South Wales* **57**: 77–91.

Fuller, M. E. 1934. The insect inhabitants of carrion, a study in animal ecology. *Bulletin of the Council for Scientific and Industrial Research, Melbourne* **82**: 5–62.

Garnett, W. B. & Foot, B. A. 1967. Biology and immature stages of *Pseudoleria crassata* (Diptera: Heleomyzidae). *Annals of the Entomological Society of America* **60**: 126–134.

Gilbert, P. & Hamilton, C. J. 1983. *Entomology: a Guide to Information Sources.* viii + 237 pp, Mansell, London.

Gill, G. D. 1962. The Heleomyzid flies of America north of Mexico. *Proceedings of the United States National Museum* **113**: 495–603.

Glaister, J. & Brash, J. C. 1937. *Medico-Legal Aspects of the Ruxton Case* (Appendix VI). Livingstone, Edinburgh & London.

Goddard, W. H. 1938. The description of the puparia of fourteen British species of Sphaeroceridae (Borboridae, Diptera). *Transactions of the Society of British Entomology* **5**: 235–258.

Gonzales, T. A., Vance, M., Helpern, M. & Umberger, C. J. 1954. *Legal Medicine: Pathology and Toxicology*, 2nd edn. 1349 pp, Appleton-Century-Crofts, New York.

Graham-Smith, G. S. 1916. Observations on the habits and parasites of common flies. *Parasitology* **8**: 440–544.

Graham-Smith, G. S. 1919. Further observations on the habits and parasites of common flies. *Parasitology* **11**: 347–384.

Grassé, P.-P. 1949–1951. *Traité de Zoologie, Insecta* Vols 9 & 10 (1–2). Masson & Cie, Paris.

Green, A. A. 1951. The control of blowflies infesting slaughterhouses. 1. Field observations on the habits of blowflies. *Annals of Applied Biology* **38**: 475–494.

Green, C. T. 1925. The puparia and larvae of sarcophagid flies. *Proceedings of the United States National Museum* **66**(29): 1–26, 9 pls.
Greenberg, B. 1971–1973. *Flies and Disease.* 2 vols, viii + 856 pp; x + 447 pp, Princeton University Press, Princeton.
Greenberg, B. 1985. Forensic Entomology: case studies. *Bulletin of the Entomological Society of America* **31**(4): 25–28.
Greenberg, B. & Szyska, M. L. 1984. Immature stages and biology of fifteen species of Peruvian Calliphoridae (Diptera). *Annals of the Entomological Society of America* **77**: 488–517.
Greig, J. 1982. Garderobes, sewers, cesspits and latrines. *Current Archaeology* (85) **8**(2): 49–52.
Grodowitz, M. M., Krchma, J. & Broce, A. B. 1982. A method of preparing soft bodied larval Diptera for scanning electron microscopy. *Journal of the Kansas Entomological Society* **55**: 751–753.
Guiart, J. 1898. Notices biographiques II. – Francesco Redi 1626–1698. *Archives de Parasitologie* **1**: 420.
Guimarães, J. H., do Prado, A. P. & Linhares, A. X. 1978. Three newly introduced blowfly species in Southern Brazil (Diptera, Calliphoridae). *Revista Brasileira Entomologia* **22**: 53–60.
Hafez, M. 1940. A study of the morphology and life-history of *Sarcophaga facultata* Pand. *Bulletin de la Société Fouad Ier d'Entomologie* **24**: 183–212.
Hall, D. G. 1948. *The Blowflies of North America.* 477 pp, [dated 1947 on cover but 1948 on title page] Say, Baltimore.
Hall, D. W. & Howe, R. W. 1953. A revised key to the larvae of the Ptinidae associated with stored products. *Bulletin of Entomological Research* **44**: 85–96.
Halstead, D. G. H. 1963. Histeroidea. *Handbooks for the Identification of British Insects* **4**(10): 1–16.
Hammer, O. 1941. Biological and ecological investigations of flies associated with pasturing cattle and their excrement. *Videnskabelige Meddelelser fra Dansk Naturhistorisk Forening i Kjøbenhavn* **105**: 1–257.
Hancox, M. 1979. The importance of non-vertebrate agents in the decomposition of sheep carrion on moorland in West Scotland. *The Western Naturalist.* **8**: 69–74.
Hanski, I. 1976a. Breeding experiments with carrion flies (Diptera) in natural conditions. *Annales Entomologici Fennici* **42**: 113–121.
Hanski, I. 1976b. Assimilation by *Lucilia illustris* (Diptera) larvae in constant and changing temperatures. *Oikos* **27**: 288–299.
Hanski, I. 1977a. Biogeography and ecology of carrion flies in the Canary Islands. *Annales Entomologici Fennici* **43**: 101–107.
Hanski, I. 1977b. An interpolation model of assimilation by larvae of the blowfly *Lucilia illustris* (Calliphoridae) in changing temperature. *Oikos* **28**: 187–195.
Hanski, I. & Kuusela, S. 1977. An experiment on competition and diversity in the carrion fly community. *Annales Entomologici Fennici* **43**: 108–115.
Hanski, I. & Nuorteva, P. 1975. Trap survey of flies and their diel periodicity in the subarctic Kevo Nature Reserve, Northern Finland. *Annales Entomologici Fennici* **43**: 56–64.
Harris, W. V. 1971. *Termites, Their Recognition and Control,* 2nd Edn. xiii + 186 pp, Longmans, New York.
Hartley, J. C. 1961. A taxonomic account of the larvae of some British Syrphidae. *Proceedings of the Zoological Society of London* **136**: 505–573.
Haskell, P. T. (Ed.) 1966. Insect Behaviour. *Symposia of the Royal Entomological Society of London* **3**: 1–113.
Hennig, W. 1943. Piophilidae. *Die Fliegen der Palaearktischen Region* **5**(40): 1–52 [in German].
Hennig, W. 1948–1952. *Die Larvenformen der Dipteren.* 3 vols. Akademie-Verlag, Berlin.
Hennig, W. 1950. Entomologische Beobachtungen an kleinen wirbeltierleichen. *Zeitschrift für hygienische Zoologie* **38**: 33–88.
Hennig, W. 1956. Beitrag zur Kenntnis der Milichiiden-Larven. *Beiträge zur Entomologie* **6**: 138–145.
Hennig, W. 1973. Diptera (Zweiflügler). *Handbuch der Zoologie, Berlin* **4**(2) (31): 1–337.
Hepburn, G. A. 1943. Sheep blowfly research. V. – Carcasses as sources of blowflies. *Journal of Veterinary Science and Animal Industry* **18**: 59–72.
Herms, W. B. 1907. An ecological and experimental study of Sarcophagidae with relation to lake beach debris. *Journal of Experimental Zoology* **4**: 45–83.
Hewitt, C. G. 1914. *The House-Fly Musca domestica Linn., Its Structure, Habits, Development, Relation to Disease and Control.* 382 pp, 104 figs, Cambridge University Press, Cambridge.
Hewson, R. & Kolb, H. H. 1976. Scavenging on sheep carcases by foxes (*Vulpes vulpes*) and badgers (*Meles meles*). *Journal of Zoology, London* **180**: 496–498.
Heymons, R., Lengerken, H. V. & Bayern, M. 1926. Studien über die Lebenserscheinungen der Silphini (Coleopt.). *Silpha obscura* L. *Zeitschrift für Morphologie und Okologie der Tiere* **6**: 287–332.
Hinton, H. E. 1945a. The Histeridae associated with stored products. *Bulletin of Entomological Research* **35**: 309–340.
Hinton, H. E. 1945b. The species of *Anthrenus* that have been found in Britain, with a description of a recently introduced species (Coleoptera, Dermestidae). *Entomologist* **78**: 6–9
Hinton, H. E. 1945c. *A Monograph of the Beetles Associated with Stored Products*, Vol. 1. 443 pp, 505 figs,

British Museum (Natural History), London.

Hinton, H. E. 1956. The larvae of the species of Tineidae of economic importance. *Bulletin of Entomological Research* **47**: 251–346.

Hinton, H. E. & Corbet, A. S. 1975. *Common Insect Pests of Stored Food Products. A Guide to Their Identification*, 5th edn. 62 pp, British Museum (Natural History), London. [Revd edn 1980, P. Freeman (Ed.)].

Hobson, R. P. 1932. Studies on the nutrition of blow-fly larvae. III. The liquefaction of muscle. *Journal of Experimental Biology* **9**: 359–365.

Hoffman, R. L. & Payne, J. A. 1969. Diplopods as carnivores. *Ecology* **50**(6): 1096–1098.

Holdaway, F. G. & Evans, A. C. 1930. Parasitism a stimulus to pupation: *Alysia manducator* in relation to the host *Lucilia sericata*. *Nature, London.* **125**: 598.

Holdaway, F. G. & Smith, H. F. 1932. A relation between size of host puparia and sex ratio of *Alysia manducator* Pantzer [sic]. *Australian Journal of Experimental Biology & Medical Science* **10**: 247–259.

Hollis, D. (Ed.) 1980 *Animal Identification, a Reference Guide.* Vol. 3. Insects. viii + 160 pp, British Museum (Natural History), London; John Wiley & Sons, Chichester, etc.

Holzer, F. J. 1939. Zerstörung an Wasserleichen durch Larven des Köcherfliege. *Zeitschrift für die gesamte gerichtliche Medizin* **31**: 223–228.

Hopkins, G. H. E. 1944. Notes on myiasis especially in Uganda. *East African Medical Journal* **21**: 258–265.

Howard, L. O. 1900. A contribution to the study of the insect fauna of human excrement. *Proceedings of the Washington Academy of Sciences* **2**: 541–604.

Howard, L. O. 1912. *The House Fly Disease Carrier, an Account of its Dangerous Activities and of the Means of Destroying it.* 312 pp. John Murray, London.

Howden, A. T. 1950. *The Succession of Beetles on Carrion.* Unpublished dissertation for degree of MS, North Carolina State College, Raleigh.

Hudson, H. F. 1914. *Lucilia sericata* Meigen attacking a live calf. *The Canadian Entomologist* **46**: 416.

Hughes, A. M. 1961. The mites of stored food. *Technical Bulletin of the Ministry of Agriculture, Fisheries and Food* **9**: 1–287.

Illingworth, J. F. 1927. Insects attracted to carrion in Southern California. *Proceedings of the Hawaii Entomological Society* **6**: 397–401.

Imms, A. D. 1939. Dipterous larvae and wound treatment. *Nature, London* **144**: 516.

Imms, A. D. 1971. *Insect Natural History*, 3rd Edn. G. C. Varley & B. M. Hobby (Eds), xviii + 317 pp, Collins, London. [New Naturalist No. 8].

Imms, A. D. 1977. *A General Textbook of Entomology*, 11th edn. (revised by O. W. Richards & R. G. Davies), 2 vols, viii + 418 pp; 936 pp, Methuen, London.

Ishijima, H. 1967. Revision of the third stage larvae of synanthropic flies of Japan (Diptera: Anthomyiidae, Muscidae, Calliphoridae and Sarcophagidae). *Japanese Journal of Sanitary Zoology* **18**: 47–100.

James, M. T. 1947. The flies that cause myiasis in man. *Miscellaneous Publications of the United States Department of Agriculture* **631**: 1–175, 98 figs.

James, M. T. 1955. The blowflies of California (Diptera: Calliphoridae). *Bulletin of the California Insect Survey* **4**: 1–34.

Jirón, L. F. 1979. Sobre moscas Califóridas de Costa Rica. *Brenesia* **16**: 221–222.

Jirón, L. F. & Cartín, V. M. 1981. Insect succession in the decomposition of the mammal in Costa Rica. *Journal of the New York Entomological Society* **89**: 158–165.

Johnson, M. D. 1975. Season and microseral variations in the insect population on carrion. *American Midland Naturalist* **93**: 79–90.

Johnston, W. & Villeneuve, G. 1897. On the medico-legal application of entomology. *Montreal Medical Journal* **26**: 81–90.

Joy, N. H. 1932. *A Practical Handbook of British Beetles.* 2 vols, xxvii + 622 pp. Witherby, London. [Reprinted 1976, Classey, Faringdon.]

Joyce, C. 1984. The detective from the laboratory. *New Scientist* **1430**: 12–16.

Kamal, A. S. 1958. Comparative study of thirteen species of Sarcosaprophagous Calliphoridae and Sarcophagidae (Diptera). 1. Bionomics. *Annals of the Entomological Society of America* **51**: 261–271.

Kano, R. 1958. Notes on flies of medical importance in Japan. Part XIV. Descriptions of five species belonging to Chrysomyinae (Calliphoridae) including one newly found species. *Bulletin of Tokyo Medical & Dental University* **5**: 465–474.

Kano, R. 1966. Estimation of the time of death on basis of fly larvae occurring on the corpse. *Tokyo Forensic Society Publications* (33rd meeting of Legal Medicine in Kanto District – special lecture) **33**: 301–302 [in Japanese].

Kano, R. & Sato, K. 1952. Notes on flies of medical importance in Japan (Part VI) larvae of Lucilini in Japan. *Japanese Journal of Experimental Medicine* **22**: 33–42.

Kano, R. & Shinonaga, S. 1968. *Fauna Japonica. Calliphoridae (Insecta: Diptera).* 181 pp, Biogeographical Society of Japan, Tokyo.

Kasule, F. K. 1968. The larval characters of some subfamilies of British Staphylinidae (Coleoptera) with keys to the known genera. *Transactions of the Royal Entomological Society of London* **120**: 115–138.
Kaufmann, R. U. 1937. Investigation on beetles associated with carrion in Pannal Ash, near Harrogate. *Entomologist's Monthly Magazine* **73**: 78–81, 227–233, 268–272.
Kaufmann, R. U. 1941. British carrion beetles. *Naturalist, Hull* **788**: 63–72; **790**: 115–124; **791** 133–138; **792**: 149–156.
Keh, B. 1985. Scope and applications of forensic entomology. *Annual Review of Entomology* **30**: 137–154.
Keilin, D. 1917. Recherches sur les Anthomyides a larves carnivores. *Parasitology* **9**: 325–450, 11 pls.
Keilin, D. 1919. Supplementary notes on the formation of a cocoon by cyclorhaphous Dipterous larvae. *Parasitology* **11**: 237–238.
Keilin, D. 1944. Respiratory systems and respiratory adaptations in larvae and pupae of Diptera. *Parasitology* **36**: 1–66
Keilin, D. & Tate, P. 1930. On certain semi-carnivorous anthomyid larvae. *Parasitology* **22**: 168–181.
Keilin, D. & Tate, P. 1940. The early stages of the families Trichoceridae and Anisopodidae (= Rhyphidae) (Diptera: Nematocera). *Transactions of the Royal Entomological Society of London* **90**: 39–62.
Kerrich, G. J., Hawksworth, D. L. & Sims, R. W. (Eds) 1978. *Key Works to the Fauna and Flora of the British Isles and Northwestern Europe*. Systematics Association Special Volume 9, xii + 179 pp, Academic Press, London, etc.
Kilpatrick, J. W. & Schoof, H. F. 1959. Interrelationship of water and *Hermetia illucens* breeding to *Musca domestica* production in human excrement. *American Journal of Tropical Medicine and Hygiene* **8**(5): 597–602.
Kitching, R. L. 1976. The immature stages of the old-world screw-worm fly, *Chrysomya bezziana* Villeneuve, with comparative notes on other Australasian species of *Chrysomya* (Diptera, Calliphoridae). *Bulletin of Entomological Research* **66**: 195–203.
Kitching, R. L. & Roberts, J. A. 1975. Laboratory observations on the teneral period in sheep blowflies, *Lucilia cuprina* (Diptera: Calliphoridae). *Entomologia Experimentalis et Applicata* **18**: 220–225.
Kitching, R. L. & Voeten, R. 1977. The larvae of *Chrysomya incisuralis* (Macquart) and *Ch. (Eucompsomyia) semimetallica* (Malloch) (Diptera: Calliphoridae). *Journal of the Australian Entomological Society* **16**: 185–190.
Klausnitzer, B. 1978. *Ordnung Coleoptera (Larven) Bestimmungs bücher zur Bodenfauna Europas*. 400 pp. 1000 figs, Junk, The Hague.
Knipling, E. B. 1958. The thermal death points of several species of insects. *Journal of Economic Entomology* **51**: 344–346.
Knipling, E. B. & Sullivan, W. N. 1957. Insect mortality at low temperatures. *Journal of Economic Entomology* **50**: 368–369.
Knipling, E. F. 1936. A comparative study of the first-instar larvae of the genus *Sarcophaga* (Calliphoridae, Diptera), with notes on the biology. *Journal of Parasitology* **22**: 417–454.
Kükenthal, W. 1968. *Handbuch der Zoologie* 4(2) (Insecta) rev. edn, M. Beier, (Ed.) De Gruyter, Berlin.
Kumar, R. & Lloyd, J. E. 1976. A bibliography of the arthropods associated with dung. *Science Series of Colorado State University Range Science Department* **18**: 1–33.
Kurahashi, H. 1971. The tribe Calliphorini from Australian and Oriental regions. 11. Calliphorini-group (Diptera: Calliphoridae). *Pacific Insects* **13**: 141–204.
Kuusela, S. & Hanski, I. 1982. The structure of carrion fly communities: the size and the type of carrion. *Holarctic Ecology* **5**: 337–348.
Laake, E. W., Cushing, E. C. & Parish, H. E. 1936. Biology of the primary screw worm fly, *Cochliomyia americana*, and a comparison of its stages with those of *C. macellaria*. *Technical Bulletin of the United States Department of Agriculture* **500**: 1–24.
Lane, R. P. 1975. An investigation into blowfly (Diptera: Calliphoridae) succession on corpses. *Journal of Natural History* **9**: 581–588.
Lane, R. P. 1978. The Diptera of Lundy Island. *Report of the Lundy Field Society* **1977**: 15–31.
Laurence, B. R. 1955. The ecology of some British Sphaeroceridae (Borboridae, Diptera). *Journal of Animal Ecology* **24**: 187–199.
Laurence, B. R. 1956. On the life history of *Trichocera saltator* (Harris) (Diptera, Trichoceridae). *Proceedings of the Zoological Society of London* **126**: 235–243.
Laurence, B. R. 1957. The British species of *Trichocera* (Diptera: Trichoceridae). *Proceedings of the Royal Entomological Society of London* (A) **32**: 132–138.
Leclercq, J. & Leclercq, M. 1948. Données bionomiques pour *Calliphora erythrocephala* Meigen et cas d'application à la médicine légale. *Bulletin de la Société Entomologique de France* **53**: 101–103.
Leclercq, M. 1968. Entomologie en Gerechtelijke Geneeskunde. *Tijdschrift Geneeskunde* **22**: 1193–1198 [in Flemish].
Leclercq, M. 1969. *Entomological Parasitology* 158 pp, Pergamon Press, Oxford.
Leclercq, M. 1975. Entomologie et médecine légale. Etude des insectes et acariens nécrophages pour

déterminer la date de la mort. *Spectrum* **17**(6): 1–7.
Leclercq, M. 1976. Entomologie et medecine legal: *Sarcophaga argyrostoma* Rob.-Desv. (Dipt., Sarcophagidae) et *Phaenica sericata* Meig. (Dipt., Calliphoridae). *Bulletin et Annales de la Société Royal Entomologique de Belgique* **112**: 119–126.
Leclercq, M. 1978. *Entomologie et Médecine Legale Datation de la Mort.* 100 pp, 14 figs, 7 pls, Masson, Paris, etc.
Leclercq, M. & Tinant-Dubois, J. 1973. Entomologie et médecine légale. Observations inédites. *Bulletin Médecine Légale Toxicologie* **16**: 251–267.
Leclercq, M. & Watrin, P. 1973. Entomologie et medecine legale: Acariens et insectes trouvés sur un cadavre humain, en décembre 1971. *Bulletin et Annales da la Société Royale Entomologique de Belgique* **109**: 195–201.
Lehrer, A. Z. 1972. Insecta Diptera Fam. Calliphoridae. *Fauna Republicii Populare Romine* **11**(12): 1–251 [in Romanian].
Levinson, Z. H. 1960. Food of housefly larvae. *Nature, London* **188**: 427–428.
Lewis, T. & Taylor, L. R. 1965. Diurnal periodicity of flight by insects. *Transactions of the Royal Entomological Society of London* **116**: 393–479.
Lindquist, A. W. 1954. Flies attracted to decomposing liver in Lake County, California. *The Pan-Pacific Entomologist* **20**: 147–152.
Lindroth, C. H. 1974. Carabidae. *Handbooks for the Identification of British Insects* **4**(2): 1–148.
Linnaeus, C. 1767. *Systema naturae, per regna tria naturae, secundum classes, ordines, genera, species, cum caracteribus, differentiis, synonymis, locis* Ed 13 rev. 1(2): 533–1327 [see p. 990].
Linssen, E. F. 1959. *Beetles of the British Isles.* 2 vols, 595 pp, 71 figs, 126 pls, Warne, London.
Lobanov, A. M. 1968 [The morphology of third instar larvae of synanthropic flies of the genus *Hydrotaea* R.-D. (Diptera, Muscidae).] *Zoologicheskii Zhurnal* **47**: 85–90 [in Russian].
Lobanov, A. M. 1970. [Morphology of the mature larvae of Helomyzidae (Diptera).] *Zoologicheskii Zhurnal* **49**: 1671–1675 [in Russian].
Lodge, O. C. 1916. Fly investigations reports. IV. Some enquiry into the question of baits and poisons for flies, being a report on the experimental work carried out during 1915 for the Zoological Society of London. *Proceedings of the Zoological Society of London* **1916**: 481–518.
Lopatenok, A. A., Boiko, L. P. & Budjakov, O. S. 1964. [Forensic significance of observations about the fauna of dead human bodies.] *Sudebno-meditsinskaya ékspertisa* **7**: 47–50 [in Russian].
Lopes, H. de Souza. 1959. A revision of Australian Sarcophagidae (Diptera). *Studia Entomologica* **2**: 33–67.
Lopes, H. de Souza 1969. Family Sarcophagidae in *A Catalogue of the Diptera of the Americas South of the United States* **103**: 1–87. Sao Paulo.
López, L. & Gisbert, J. A. 1962. *Tratado de Medicina Legal* 827 pp, Saber, Valencia.
Lord, W. D. & Burger, J. F. 1983. Collection and preservation of forensically important entomological materials. *Journal of Forensic Sciences* **28**: 936–944.
Lord, W. D. & Burger, J. F. 1984. Arthropods associated with harbor seal (*Phoca vitulina*) carcasses stranded on islands along the New England coast. *International Journal of Entomology* **26**: 282–285.
Lothe, F. 1964. The use of larva infestation in determining the time of death. *Medicine Science and the Law* **4**: 113–115.
Lowne, B. T. 1890–1895. *The Anatomy, Physiology, Morphology and Development of the Blow-fly (Calliphora erythrocephala). A Study in the Comparative Anatomy and Morphology of Insects.* 2 vols 1–350 pp + 351–778 pp, 108 figs, 52 pls, Porter, London.
Lundt, H. 1964. Ecological observations about the invasion of insects into carcasses buried in soil. *Pedobiologia* **4**: 158–180 [in German with English summary].
Luther, H. 1946. Beobachtungen über *Phallus impudicus* (L.) Pers. in Finland. *Memoranda Societatis pro Fauna et Flora Fennica* **23**: 42–59.
Lyneborg, L. 1970. Taxonomy of European *Fannia* larvae (Diptera, Fanniidae). *Stuttgarter Beiträge zur Naturkunde* **215**: 1–28.
McAlpine, J. F. 1965. Insects and related terrestrial invertebrates of Ellef Ringnes Island. *Arctic* **18**: 73–103.
McAlpine, J. F. 1977. A revised classification of the Piophilidae, including 'Neottiophilidae' and 'Thyreophoridae' (Diptera: Schizophora). *Memoirs of the Entomological Society of Canada* **103**: 1–66.
McAlpine, J. F., Peterson, B. V., Shewell, G. E., Teskey, H. J., Vockeroth, J. R. & Wood, D. M. 1981. *Manual of Nearctic Diptera* Vol. 1. 674 pp, Agriculture Canada Monograph No. 27 (Quebec).
Mackerras, I. M. (Ed.) 1970. *The Insects of Australia* xiii + 1029 pp, CSIRO, Melbourne.
Mackerras, M. J. 1933. Observations on the life-histories, nutritional requirements and fecundity of blowflies. *Bulletin of Entomological Research* **24**: 353–362.
McKnight, B. E. (Transl.) 1981. *The washing away of wrongs: Forensic Medicine in thirteenth-Century China* 181 pp. University of Michigan, Ann Arbor.
MacLeod, J. & Donnelly, J. 1956. Methods for the study of blowfly populations. 1. Bait-trapping. Significance limits for comparative sampling. *Annals of Applied Biology* **44**: 80–104.

MacLeod, J. & Donnelly, J. 1957a. Individual and group marking methods for fly population studies. *Bulletin of Entomological Research* **48**: 585–592.
MacLeod, J. & Donnelly, J. 1957b. Some ecological relationships of natural populations of calliphorine blowflies. *Journal of Animal Ecology* **26**: 135–170.
MacLeod, J. & Donnelly, J. 1958. Local distribution and dispersal paths of blowflies in hill country. *Journal of Animal Ecology* **27**: 349–374.
MacLeod, J. & Donnelly, J. 1960. Natural features and blowfly movement. *Journal of Animal Ecology* **29**: 85–93.
MacLeod, J. & Donnelly, J. 1962. Microgeographic aggregations in blowfly populations. *Journal of Animal Ecology* **31**: 525–543.
MacLeod, J. & Donnelly, J. 1963. Dispersal and interspersal of blowfly populations. *Journal of Animal Ecology* **32**: 1–32.
Macquart, M. 1835. *Histoire Naturelle des Insectes. Diptères.* **2**: 1–701 [see p. 241].
Magy, H. I. & Black, R. J. 1962. An evaluation of the migration of fly larvae from garbage cans in Pasadena, California. *California Vector Views* **9**: 55–60.
Malloch, J. R. 1917. A preliminary classification of Diptera, exclusive of Pupipara, based upon larval and pupal characters, with keys to imagines in certain families. Part 1. *Bulletin of the Illinois State Laboratory of Natural History* **12**: 161–409. pls XXVIII–LVII.
Mangan, R. L. 1977. A key and selected notes for the identification of larvae of Sepsidae (Diptera) from the temperate regions of North America. *Proceedings of the Entomological Society of Washington* **79**: 338–342.
Marchenko, M. I. 1978. [Studies on the destruction of cadavers by insects]. *Sudebno-Meditsinskaya ékspertiza* **21**(1): 17–20.
Martin, J. E. H. 1978. *The Insects and arachnids of Canada. Part 1. Collecting, preparing and rearing insects, mites and spiders.* Agriculture Canada Publication 1643, 182 pp.
Meek, C. L., Andis, M. D. & Andrews, C. S. 1983. Role of the entomologist in forensic pathology, including a selected bibliography. *Bibliographies of the Entomological Society of America* **1**: 1–10.
Mégnin, P. 1887. La Faune des Tombeaux. *Compte Rendu Hebdomadaire des Séances de l'Académie des Sciences*, Paris **105**: 948–951. (also *C. R. Mem. Soc. Biol. Paris* (8) **4**(39): 655–658; and 1888. *Ann. Hyg. Méd. Paris* (3) **19**: 160–166).
Mégnin, P. 1894. *La Faune des Cadavres.* Encyclopédie Scientifique des Aide-Memoire, 214 pp, 28 figs, G. Masson, Gauthier-Villars et Fils, Paris.
Mellanby, K. 1938. Diapause and metamorphosis of the blowfly *Lucilia sericata* Meig. *Parasitology* **30**: 392–399.
Mellanby, K. 1943. *Scabies* 81 pp, Oxford University Press [Reprinted 1972, E. W. Classey Ltd, Faringdon].
Mengyu, Z. 1982. A study of the larvae of some common sarcophagid flies from China. *Entomotaxonomia* **4**: 93–106 [in Chinese with English summary].
Michelsen, V. 1983. *Thyreophora anthropophaga* Robineau-Desvoidy, an 'extinct' bone-skipper rediscovered in Kashmir (Diptera: Piophilidae, Thyreophorina). *Entomologica Scandinavica* **14**: 411–414.
Mihályi, F. 1967. Seasonal distribution of the synanthropic flies in Hungary. *Annales Historico-Naturales Musei Nationalis Hungarici Pars Zoologica* **59**: 327–344.
Moore, B. P. 1955. Notes on carrion coleoptera in the Oxford district. *Entomologist's Monthly Magazine* **91**: 292–295.
Morley, C. 1907. Ten years' work among vertebrate carrion. *Entomologist's Monthly Magazine* **43**: 45–51.
Motter, M. G. 1898. A contribution to the study of the fauna of the grave. A study of one hundred and fifty disinterments, with some additional experimental observations. *Journal of the New York Entomological Society* **6**: 201–231.
Müller, B. 1975. *Gerichtliche Medizin,* 2nd edn. 2 vols, 1331 pp, Springer-Verlag, Berlin. [First pub. 1953]
Nabaglo, L. 1973. Participation of invertebrates in decomposition of rodent carcasses in forest ecosystems. *Ekologia Polska* **21**: 251–270.
Nagasawa, S. & Kishino, M. 1965. Application of Pradhan's formula to the pupal development of the common house fly *Musca domestica vicina* Macquart. *Japanese Journal of Applied Entomology and Zoology* **9**: 94–98.
Nandi, B. C. 1980. Studies on the larvae of flesh flies from India (Diptera: Sarcophagidae). *Oriental Insects* **14**: 303–323.
Nielsen, B. O. 1963. Om forekomsten af Diptera på Alm. Stinksvamp (*Phallus impudicus* Pers.) og fund af *Phaonia errans* Mg. og *Helomyza fuscicornis* Zett., nye for den danske fauna. *Flora og Fauna*, Silkeborg **69**: 126–134. [English summary].
Nielsen, B. O. & Nielsen, S. A. 1946. Schmeissfliegen (Calliphoridae) und vakuumverpackter Schinken. *Anzeiger für Schadlingskunde Pflanzen-und Umweltschutz.* **49**: 113–115.
Nixon, G. [E. J.] 1954. *The World of Bees.* 214 pp, 16 figs, Hutchinson, London.
Norn, M. S. 1971. *Demodex folliculorum.* Incidence, regional distribution, pathogenicity. *Danish Medical Bulletin.* **18**: 14–17.

Norris, K. R. 1957. A method of marking Calliphoridae (Diptera) during emergence from the puparium. *Nature, London* **180**: 1002.
Norris, K. R. 1959. The ecology of sheep blowflies in Australia. *In* Keast, A., Crocker, R. L. & Christian, C. S. (Eds) Biogeography and Ecology in Australia. *Monographia Biologicae* **8**: 514–544.
Norris, K. R. 1965a. The bionomics of blow flies. *Annual Review of Entomology* **10**: 47–68.
Norris, K. R. 1965b. Daily patterns of flight activity of blowflies (Calliphoridae: Diptera) in the Canberra district as indicated by trap catches. *Australian Journal of Zoology* **14**: 835–853.
Norris, K. R. [in press]. [Revision of Australian Calliphoridae].
Norris, K. R. & Upton, M. S. 1974. The collection and preservation of insects. *Australian Entomological Society Miscellaneous Publications* No. 3 (2nd Edn), 33 pp.
Nuorteva, P. 1952. Die Nahrungspflanzenwahl der Insekten im Lichte von untersuchungen an Zikaden. *Annales Academiae Scientiarum Fennica (Suomalaisen Tiedeakatemian Toimituksia)* (A) IV Biol. **19**: 1–90.
Nuorteva, P. 1958. Some peculiarities in the seasonal occurrence of poliomyelitis in Finland. *Annales medicinae experimentalis et Biologiae Fenniae* **36**: 335–342.
Nuorteva, P. 1959a. Studies on the significance of flies in the transmission of poliomyelitis I. The occurrence of *Lucilia* species (Diptera, Calliphoridae) in relation to the occurrence of poliomyelitis in Finland. *Annales Entomologici Fennici* **25**: 1–24.
Nuorteva, P. 1959b. Studies on the significance of flies in the transmission of poliomyelitis. III. The composition of the blowfly fauna and the activity of the flies during the epidemic season of poliomyelitis in South Finland. *Annales Entomologici Fennici* **25**: 121–136.
Nuorteva, P. 1959c. Studies on the significance of flies in the transmission of poliomyelitis. IV. The composition of the blowfly fauna in different parts of Finland during 1958. *Annales Entomologici Fennici* **25**: 137–162.
Nuorteva, P. 1961a. Studies on the significance of flies in the transmission of poliomyelitis. VII. Attraction of blowflies to a house by the honeydew of the aphid *Phorodon humuli* (Schrk.). *Annales Entomologici Fennici* **27**: 51–53.
Nuorteva, P. 1961b. Liero-ja raatokärpästen jatkuva tarkkailu aloitettu Urjalassa. *Lounais-Hämeen Luonto* **11**: 75–82.
Nuorteva, P. 1963. Synanthropy of blowflies (Dipt., Calliphoridae) in Finland. *Annales Entomologici Fennici* **29**: 1–49.
Nuorteva, P. 1964. Differences in the ecology of *Lucilia caesar* (L.) and *Lucilia illustris* (Meig.) (Diptera, Caliphoridae) in Finland. *Wiadomosci Parazytologiczne* **10**: 583–587.
Nuorteva, P. 1965. The flying activity of blowflies (Dipt., Calliphoridae) in subarctic conditions. *Annales Entomologici Fennici* **31**: 242–245.
Nuorteva, P. 1966a. The flying activity of *Phormia terrae-novae* R.-D. (Dipt., Calliphoridae) in subarctic conditions. *Annales Zoologici Fennici* **3**: 73–81.
Nuorteva, P. 1966b. Local distribution of blowflies in relation to human settlement in an area around the town of Forssa in South Finland. *Annales Entomologici Fennici* **32**: 128–137.
Nuorteva, P. 1966c. The occurrence of *Phormia terrae-novae* R.-D. (Dipt., Calliphoridae) and other blowflies in the archipelago of the subarctic Lake Inarinjärvi. *Annales Entomologici Fennici* **32**: 240–251.
Nuorteva, P. 1970. Histerid beetles as predators of blowflies (Diptera, Calliphoridae) in Finland. *Annales Zoologici Fennici* **7**: 195–198.
Nuorteva, P. 1971. Methylquecksilber in den Nahrungsketten dur Nature. *Naturwissenschaftliche Rundschau Stuttgart* **24**: 233–243.
Nuorteva, P. 1972. A three year study of the duration of development of *Cynomyia mortuorum* (L.) (Dipt., Calliphoridae) in the conditions of a subarctic fell. *Annales Entomologici Fennici* **38**: 65–74.
Nuorteva, P. 1974. Age determination of a blood stain in a decaying shirt by entomological means. *Forensic Science* **3**: 89–94.
Nuorteva, P. 1977. Sarcosaprophagous insects as forensic indicators. *In* Tedeschi, C. G., Eckert, W. G. & Tedeschi, L. G. *Forensic Medicine, a Study in Trauma and Environmental Hazards*. Vol. II. *Physical Trauma*. pp. 1072–1095, Saunders, Philadelphia, etc.
Nuorteva, P. & Häsänen, E. 1972. Transfer of mercury from fishes to sarcosaprophagous flies. *Annales Zoologici Fennici* **9**: 23–27.
Nuorteva, P. & Laurikainen, E. 1964. Synanthropy of blowflies (Dipt., Calliphoridae) on the island of Gotland, Sweden. *Annales Entomologici Fennici* **30**: 187–190.
Nuorteva, P., Isokoski, M. & Laiho, K. 1967. Studies on the possibilities of using blowflies (Dipt.) as medico-legal indicators in Finland. *Annales Entomologici Fennici* **33**: 217–225.
Nuorteva, P., Kotimaa, T., Pohjolainen, L. & Räsänen, T. 1964. Blowflies (Dipt., Calliphoridae) on the refuse depot of the city of Kuopio in central Finland. *Annales Entomologici Fennici* **30**: 94–104.
Nuorteva, P., Schumann, H., Isokoski, M. & Laiho, K. 1974. Studies on the possibilities of using blowflies (Dipt., Calliphoridae) as medico-legal indicators in Finland. 2. Four cases where species identification was performed from larvae. *Annales Entomologici Fennici* **40**(2): 70–74.

O'Flynn, M. A. 1983. The succession and rate of development of blowflies in carrion in Southern Queensland and the application of these data to forensic entomology. *Journal of the Australian Entomological Society* **22**: 137–148.
O'Flynn, M. A. & Moorehouse, D. E. 1980. Identification of early immature stages of some common Queensland carrion flies. *Journal of the Australian Entomological Society* **19**: 53–61.
Ohtaki, T. 1966. On the delayed pupation of the fleshfly *Sarcophaga peregrina* Robineau-Desvoidy. *Japanese Journal of Medical Science & Biology* **19**: 97.
Okada, T. 1968. *Systematic Study of the Early Stages of Drosophilidae.* 188 pp, 84 figs, Bunka Zugeisha, Tokyo.
Okely, E. F. 1969. *The Biology of Some British Sphaerocerid Flies.* A thesis submitted for the Diploma of Imperial College, University of London. Imperial College of Science and Technology.
Okely, E. F. 1974. Description of the puparia of twenty-three British species of Sphaeroceridae (Diptera, Acalyptratae). *Transactions of the Royal Entomological Society of London* **126**: 41–56.
Oldroyd, H. 1954. The Seaweed fly nuisance. *Discovery* **15**: 198–202.
Oldroyd, H. 1964. *The Natural History of Flies* 324 pp, 40 figs, 32 pls, Weidenfeld & Nicolson, London.
Oldroyd, H. 1970a. *Collecting, Preserving and Studying Insects,* 2nd edn. 327 pp, 15 pls, Hutchinson, London.
Oldroyd, H. 1970b. Diptera 1. Introduction and key to families. *Handbooks for the Identification of British Insects.* **9**(1): 1–104.
Oldroyd, H. & Smith, K. G. V. 1973. Eggs and larvae of flies. *In* Smith, K. G. V. (Ed.) *Insects and Other Arthropods of Medical Importance.* pp. 289–323, British Museum (Natural History), London.
Osten Sacken, C. R. 1887. On Mr. Portchinski's publications on the larvae of Muscidae including a detailed abstract of his last paper: comparative biology of the necrophagous and coprophagous larvae. *Berliner Entomologischer Zeitschrift* **31**: 17–28.
Osten Sacken, C. R. 1894. *On the oxen-born bees of the ancients (Bugonia), and their relation to Eristalis tenax, a two-winged insect.* (Privately published) xiv + 80 pp. Heidelberg. [A privately published supplement appeared in 1895, also Heidelberg. A shorter account was first published in *Bolletino della Società Entomologica Italiana* in 1893.]
Ozerov, A. L. 1984. The biology of *Protothyreophora grunini* Ozerov (Diptera, Thyreophoridae) *Nauchnye Doklady Vysshei Shkoly. Biologicheskie Nauki* **1984** (4) (244): 39–41.
Papavero, N. (Ed.) 1966 – [current]. *A Catalogue of the Diptera of the Americas South of the United States.* Departmento de Zoologia, Sao Paulo. [Appears in fascicles, one per family, each with a bibliography; over 80 (of 110 intended) published so far]
Papp, L. & Plachter, H. 1976. On cave-dwelling Sphaeroceridae from Hungary and Germany (Diptera). *Annales Historico-Naturales Musei Nationalis Hungarici* **68**: 195–207.
Paramonov, S. J. 1954. Notes on Australian Diptera (XIII-XV) – [XV. On the family Thyreophoridae (Acalyptrata)]. *Annals and Magazine of Natural History* (12) **7**: 275–297. [Thyreophoridae pp. 292–297.]
Parish, H. E. & Cushing, E. C. 1938. Locations for blowfly traps: abundance and activity of blowflies and other flies in Menard County, Texas. *Journal of Economic Entomology* **31**: 750–763.
Parmenter, L. 1952. *Leptocera (Heteroptera) ferruginata* Stenhammar (Dipt., Sphaeroceridae) on a dead mole in Surrey. *Entomologist's Monthly Magazine* **88**: 63.
Patton, W. S. 1922. Some notes on Indian Calliphorini IV. *Chrysomyia albiceps* Wied. (*rufifacies* Froggatt); one of the Australian sheep maggot flies and *Chrysomyia villeneuvii* sp. n. *Indian Journal of Medical Research* **9**: 561–569.
Patton, W. S. 1933. Studies on the higher Diptera of medical and veterinary importance. A revision of the species of the genus *Musca*, based on a comparative study of the male terminalia. II. A practical guide to the Palaearctic species. *Annals of Tropical Medicine and Parasitology* **27**: 397–430.
Pavillard, E. R. & Wright, E. A. 1957. An antibiotic from maggots. *Nature, London* **180**: 916–917.
Payne, J. A. 1965. A summer carrion study of the baby pig *Sus scrofa* Linnaeus. *Ecology* **46**: 592–602.
Payne, J. A. & Crossley, D. A. 1966. Animal species associated with pig carrion. Health Phys. Div. A. Programme Report ORNL-TM-1432, 70 pp, Oak Ridge National Laboratory, Tennessee.
Payne, J. A. & King, E. W. 1969. Lepidoptera associated with pig carrion. *Journal of the Lepidopterists' Society* **23**(3): 191–195.
Payne, J. A. & King, E. W. 1970. Coleoptera associated with pig carrion. *Entomologist's Monthly Magazine* **105**(1969): 224–232.
Payne, J. A. & King, E. W. 1972. Insect succession and decomposition of pig carcasses in water. *Journal of the Georgia Entomological Society* **7**(3): 153–162.
Payne, J. A. & Mason, W. R. M. 1971. Hymenoptera associated with pig carrion. *Proceedings of the Entomological Society of Washington* **73**: 132–141.
Payne, J. A., King, E. W. & Beinhart, G. 1968a. Arthropod succession and decomposition of buried pigs. *Nature, London* **219**(5159): 1180–1181.
Payne, J. A., Mead, F. W. & King, E. W. 1968b. Hemiptera associated with pig carrion. *Annals of the Entomological Society of America* **61**(3): 565–567.

Peacock, E. R. 1977. Rhizophagidae. *Handbooks for the Identification of British Insects* **5**(5a): 1–19.
Pessoa, S. P. & Lane, P. 1941. Coléopteros necrôfageos de interêsse médico-legal. Ensáio monográfico sôbre a familia Scarabaeidae de S. Paulo e regios uizinhas. *Archivos de Zoologia do Estado de Sao Paulo* **2**: 389–504 [in Portugese].
Pitkin, B. R. [in press] Sphaeroceridae. *Handbooks for the Identification of British Insects*
Pitkin, B. R. [in press]. Sphaeroceridae (Diptera) caught in baited traps at Silwood Park, Berks. *Ecological Entomology*.
Pont, A. C. 1979. Sepsidae, Diptera Cyclorrhapha, Acalyptrata. *Handbooks for the Identification of the British Insects* **10**(5c): 1–35.
Pont, A. C. & Matile, L. 1980. Découverte de quelques Insectes de J. P. Mégnin; identité d'*Ophyra cadaverina* Mégnin (1894). *Bulletin de la Société Entomologique de France* **85**: 41–43.
Portschinsky, I. A. 1885. Muscarum cadaverinarum stercorariarumque biologia comparata. *Horae Societatis Entomologicae Rossicae* **19**: 210–244.
Portschinsky, I. A. 1913. *Muscina stabulans* Fall., mouche nuisible à l'homme et à son ménage, en état larvaire destructeuse des larves de *Musca domestica*. *Trudý Byuro po Entomologii* **10**: 1–39 [in Russian].
Povolný, D. & Rozsypal, J. 1968. Towards the autecology of *Lucilia sericata* (Meigen, 1826) (Dipt., Call.) and the origin of its synanthropy. *Acta Scientiarum Naturalium Academiae Scientiarum Bohemoslovacae Brno* (2 N.S.) **8**: 1–32.
Pradhan, S. 1946. Insect population studies. IV. Dynamics of temperature effect on insect development. *Proceedings of the National Institute of Sciences of India* **12**: 385–404.
Prins, A. J. 1982. Morphological and biological notes on six South African blow-flies (Diptera, Calliphoridae) and their immature stages. *Annals of the South African Museum* **90**: 201–217.
Putman, R. J. 1977. Dynamics of the blowfly, *Calliphora erythrocephala*, within carrion. *Journal of Animal Ecology* **46**: 853–866.
Putman, R. J. 1978a. Flow of energy and organic matter from a carcase during decomposition. Decomposition of small mammal carrion in temperate systems. 2. *Oikos* **31**: 58–68.
Putman, R. J. 1978b. The role of carrion-frequenting arthropods in the decay process *Ecological Entomology* **3**: 133–139.
Putman, R. J. 1983. Carrion and Dung: the decomposition of animal wastes. *Studies in Biology* Institute of Biology, London **156**: 1–62.
Ragge, D. R. 1965. *Grasshoppers, Crickets and Cockroaches of the British Isles*. xii + 299 pp, Warne, London.
Ragge, D. R. 1973. The British Orthoptera: a supplement. *Entomologist's Gazette* **24**: 227–245.
Ratcliffe, F. N. 1935. Observations on the sheep blowfly (*Lucilia sericata* Meig.) in Scotland. *Annals of Applied Biology* **22**: 742–753.
Réamur, R. A. F. de. 1738. *Mémoires pour servir à l'histoire des insectes*, **4**, Paris.
Reed, H. B. 1958. A study of dog carcase communities in Tennessee, with special reference to the insects. *The American Midland Naturalist* **59**: 213–245.
Reiter, C. 1984. Zum Wachstumsverhalten der Maden der blauen Schmeissfliege *Calliphora vicina*. *Zeitschrift für Rechtsmedizin* **91**: 295–308.
Reiter, C. & Wollenek, G. 1982. Bermerkungen zur Morphologie forensisch bedeutsamer Fliegenmaden. *Zeitschrift für Rechtsmedizen* **89**: 197–206.
Reiter, C. & Wollenek, G. 1983a. Zur Artbestimmung der Maden forensisch bedeutsamer Schmeissfliegen. *Zeitschrift für Rechtsmedizin* **90**: 309–316.
Reiter, C. & Wollenek, G. 1983b. Zur Artbestimmung der Puparien forensisch bedeutsamer Schmeissfliegen. *Zeitschrift für Rechtsmedizin* **91**: 61–69.
Remmert, H. 1965. Distribution and the ecological factors controlling distribution of the European wrackfauna. *Botanica Gothoburgensia* **3**: 179–184
Richards, O. W. 1930. The British species of Sphaeroceridae (Borboridae, Diptera). *Proceedings of the Zoological Society of London* **1930**: 261–345.
Richards, O. W. 1977. Hymenoptera. Introduction and keys to families, 2nd edn. *Handbooks for the Identification of British Insects* **6**(1): 1–100.
Roback, S. S. 1954. The evolution and taxonomy of the Sarcophaginae (Diptera, Sarcophagidae). *Illinois Biological Monographs* **23**: (3/4): 1–181, 1 fig., 34 pls.
Roberts, R. A. 1933. Activity of blowflies and associated insects at various heights above the ground. *Ecology* **14**: 306–314.
Robinson, G. S. 1979. Clothes-moths of the *Tinea pellionella* complex: a revision of the World's species (Lepidoptera: Tineidae). *Bulletin of the British Museum Natural History (Entomology)* **38**: 57–128.
Roch, M. 1948. Les Piqueres d'Hymenoptères. Traité de Médicine, IV. Masson, Paris.
Rodriguez, W. C. & Bass, W. M. 1983. Insect activity and its relationship to decay rates of human cadavers in east Tennessee. *Journal of Forensic Sciences*. **28**: 423–432.
Roháček, J. 1982. A monograph and re-classification of the previous genus *Limosina* Macquart (Diptera, Sphaeroceridae) of Europe. *Beiträge zur Entomologie* **32**: 195–282.

Rohdendorf, B. 1930–. Sarcophaginae. *Die Fliegen der Palaearktischen Region* **8**(64): 1– [still appearing].
Rohe, D. L., Magy, H. I., White, K. E. & Linsdale, D. D. 1964. An evaluation of fly larval migration from containers of combined refuse in the city of Compton. *California State Department of Public Health, Bureau of Vector Control*, 35 pp [multilithed].
Rohe, D. L., Magy, H. I., White, K. E., Linsdale, D. D. & Estes, R. J. 1963. Fly larval migration from residential garbage cans, City of Long Beach. *California State Department of Public Health, Bureau of Vector Control*, 10 pp [multilithed].
Rondani, C. 1874. Species italicae ordinis dipterorum (Muscaria Rndn.) Stirps XXII. – Lonchaeinae Rndn. collectae et observatae. *Bolletino della Società Entomologica Italiana* **6**: 243–274.
Roy, D. N. & Siddons, L. B. 1939. On the life-history and bionomics of *Chrysomyia rufifacies* Macq. (Order Diptera, Family Calliphoridae). *Parasitology* **31**: 442–447.
Rozkosný, R. 1982–1983. A Biosystematic Study of the European Stratiomyidae (Diptera), 2 vols. *Series Entomologica* **25**: 401 + 403 pp, 147 pls, 136 maps.
Rubezhanskiv, A. F. 1965. [Fly puparia as forensic indicators.] *Sudebno-meditsinskaya ékspertisa* **8**: 37 [in Russian].
Rubezhanskiv, A. F. & Ozanovskii, V. S. 1964. Forensic conclusions drawn from insects occurring on skeletonized human corpses. *Sudebno-meditsinskaya ékspertisa* **7**: 50.
Sabrosky, C. W. 1983. A synopsis of the world species of *Desmometopa* Loew (Diptera, Milichiidae). *Contributions of the American Entomological Institute* **19**(8): 1–69.
Sanjean, J. 1957. Taxonomic studies of *Sarcophaga* larvae of New York, with notes on the adults. *New York (Cornell) Agricultural Experimental Station Memoirs* **349**: 1–115, 149 figs.
Satchell, G. H. 1947. The larvae of the British species of *Psychoda* (Diptera, Psychodidae). *Parasitology* **38**: 51–59.
Schmitz, H. 1928. Occurrence of phorid flies in human corpses buried in coffins. *Natuurhistorisch Maandblad* **17**: 150.
Schumann, H. 1953–1954. Morphologisch-systematische Studien an Larven von hygienisch wichtigen mitteleuropäischen Dipteren der Familien Calliphoridae – Muscidae. *Wissenschaftliche Zeitschrift der Universität Greifswald* **3**: 245–274.
Schumann, H. 1962. Zur Morphologie einiger larven der Familien Borboridae und Sepsidae (Diptera). *Mitteilungen aus dem Zoologischen Museum in Berlin* **38**: 415–450.
Schumann, H. 1963. Zur Larvalsystematik der Muscinae nebst Beschreibung einiger Musciden- und Anthomyidenlarven. *Deutsche Entomologische Zeitschrift* N.F. **10**: 134–151, 12 pls.
Schumann, H. 1965. Merkblätter über angewandte Parasitenkunde und Schädlingsbekampfung. Merkblatt Nr. 11. Die Schmeissfliegengattung *Calliphora*. *Angewandte Parasitologie* [Suppl.] **6**(3): 1–14.
Schumann, H. 1971. Merkblätter über angewandte Parasitenkunde und Schädlingsbekampfung. Merkblatt Nr. 18. Die Gattung *Lucilia* (Goldfliegen). *Angewandte Parasitologie* [Suppl.] **12**(4): 1–20.
Séguy, E. 1923*a*. Étude sur le *Muscina stabulans* Fallén (Diptère). *Bulletin du Muséum d'Histoire Naturelle Paris* **29**: 310–317.
Séguy, E. 1923*b*. Note sur les larves des *Muscina stabulans* et *assimilis* (Diptères). *Bulletin du Muséum d'Histoire Naturelle Paris* **29**: 443–445.
Séguy, E. 1928. Études sur les mouches parasites. 1. Conopides, Oestrides et Calliphorines de l'Europe occidentale. *Encyclopédie Entomologique* (A) **9**: 1–251.
Séguy, E. 1941. Études sur les mouches parasites. 2. Calliphorides, Calliphorines (suite), Sarcophaginae et Rhinophorinae de l'Europe occidentale et méridionale. Recherches sur la morphologie et la distribution géographiques des Diptères à larves parasites. *Encyclopédie Entomologique* (A) **21**: 5–436.
Séguy, E. 1950. La Biologie des Diptères. *Encyclopedie Entomologique* **26**: 1–609.
Senior-White, R. A., Aubertin, D. & Smart, J. 1940. *The Fauna of British India including the Remainder of the Oriental Region, 6. Family Calliphoridae*. xii + 288 pp, 152 figs, Taylor & Francis, London.
Shorrocks, B. 1972. *Drosophila* [Invertebrate Types] 144 pp, 53 figs, Ginn, London.
Shubeck, P. P. 1968. Orientation of carrion beetles to carrion: random or non-random. *Journal of the New York Entomological Society* **76**: 253–265.
Simmonds, M. S. J. 1984. *Parasitoids of Synanthropic Flies*. Ph.D. thesis, Birkbeck College, University of London.
Simmons, P. 1927. The cheese skipper as a pest in cured meats. *Department Bulletin, U.S. Department of Agriculture*, **1453**: 1–55.
Simpson, K. 1980. *Forty Years of Murder* 398 pp, Granada, London, [First published by Harrap, 1978.]
Simpson, K. 1985. *Forensic Medicine*, 9th edn. 356 pp, Arnold, London.
Sims, R. W. (Ed.) 1980. *Animal Identification, A Reference Guide. Vol. 2. Land and Freshwater Invertebrates (not insects)* x + 120 pp, British Museum (Natural History), London. [see also Hollis, D. (1980).]
Skidmore, P. 1967. The biology of *Scoliocentra villosa* (Mg.) (Dipt., Heleomyzidae). *Entomologist's Monthly Magazine* **102**: 94–98.
Skidmore, P. 1973. Notes on the biology of Palaearctic Muscids (1). *Entomologist* **106**: 25–48.

Skidmore, P. 1985. The Biology of the Muscidae of the World. *Series Entomologica* **29**: xiv + 550 pp, 160 figs.
Slater, J. A. & Baranowski, R. M. 1978. *How to Know the True Bugs (Hemiptera-Heteroptera)*. x + 256 pp, Brown, Dubuque [Iowa].
Smart, J. 1935. The effects of temperature and humidity on the cheese skipper, *Piophila casei* (L.). *Journal of Experimental Biology* **12**: 384–388.
Smit, B. 1931. A study of the sheep blowflies of South Africa. *Report of the Director of Veterinary Service, Onderstepoort* **17**: 299.
Smit, F. G. A. M. 1957. Siphonaptera. *Handbooks for the Identification of British Insects*. **1**(16): 1–94, 200 figs.
Smith, C. N. 1933. Notes on the life history and molting process of *Sarcophaga securifera* Villeneuve. *Proceedings of the Entomological Society of Washington* **35**: 159–164.
Smith, K. G. V. 1956. On the Diptera associated with the stinkhorn (*Phallus impudicus* Pers.) with notes on other insects and invertebrates found on this fungus. *Proceedings of the Royal Entomological Society of London* (A) **31**: 49–55.
Smith, K. G. V. 1970. The nature and succession of the invertebrate fauna. *In* Easton, A. M. & Smith, K. G. V. The entomology of the cadaver. *Medicine, Science & the Law* **10**: 208–215, 3 pls.
Smith, K. G. V. 1973a. Insects wholemounts. *In* Gray, P. (Ed.) *The Encyclopaedia of Microscopy and Microtechnique* pp. 280–281, van Nostrand Reinhold, New York.
Smith, K. G. V. (Ed.). 1973b. *Insects and Other Arthropods of Medical Importance*. 576 pp, 12 pls, 217 figs British Museum (Natural History), London. [Forensic entomology pp. 483–486.]
Smith, K. G. V. 1974. Rearing the Hymenoptera Parasitica. *Leaflets of the Amateur Entomologist's Society* **35**: 1–15.
Smith, K. G. V. 1975. The faunal succession of insects and other invertebrates on a dead fox. *Entomologist's Gazette* **26**: 277–287.
Smith, K. G. V. (Convener) *et al.* 1976. Diptera. *In* Kloet and Hincks *A Check List of British Insects*. *Handbooks for the Identification of British Insects* **11**(5): 1–139.
Smith, K. G. V. 1981. The larva of *Dryomyza anilis* Fall (Dipt., Dryomyzidae) with a tentative key for the separation of the larvae of some superficially allied families. *Entomologist's Monthly Magazine* **116** *(1980)*: 167–170.
Smith, K. G. V. [in press] Darwin's Insects. *Bulletin of the British Museum (Natural History) Historical Series*.
Smith, K. G. V. & Dear, J. P. 1978. National fish skin week. *Antenna* **2**: 125–126
Smith, K. G. V. & Grensted, L. W. 1963. The larva of *Psychoda (Philosepedon) humeralis* Meigen (Diptera, Psychodidae). *Parasitology* **53**: 155–156.
Smith, K. G. V. & Thomas, V. 1979. Intestinal myiasis in man caused by *Clogmia* (= *Telmatoscopus*) *albipunctatus* (Psychodidae, Diptera). *Transactions of the Royal Society of Tropical Medicine & Hygiene*. **73**: 349–350.
Soós, Á. (Ed.) 1984– [still appearing]. *Catalogue of Palaearctic Diptera*. Akadémiai Kiadó & Elsevier Science Publishers, Budapest & Amsterdam. [To be published in 14 volumes, 2 per year.]
Southwood, T. R. E.. 1978. *Ecological Methods: With particular reference to the study of insect populations*. 2nd edn. 524 pp, ill., Chapman & Hall, London.
Southwood, T. R. E. & Leston, D. 1959. *Land and Water Bugs of the British Isles*. 436 pp, 153 figs, 63 pls, Warne, London.
Spector, W. 1943. Collecting beetles (*Trox*) with feather bait traps (Coleoptera: Scarabaeidae). *Entomological News* **54**: 224–229.
Springett, B. P. 1968. Aspects of the relationship between burying beetles, *Necrophorus* spp. and the mite *Poecilochirus necrophori* Vitz. *Journal of Animal Ecology* **37**: 417–424.
Stakelberg, A. A. 1956. [Diptera associated with man from the Russian fauna]. *Opredeliteli po Faune SSSR* **60**: 1–164 [in Russian].
Steele, B. F. 1927. Notes on the feeding habits of carrion beetles. *Journal of the New York Entomological Society* **35**: 77–81.
Steffan, W. A. & Evenhuis, N. L. [In preparation]. *Catalog of Australasian and Oceanian Diptera*. B. P. Bishop Museum, Honolulu. [Publication expected in 1988.]
Steyskal, G. C. 1957. The relative abundance of flies (Diptera) collected at human feces. *Zeitschrift für Angewandte Zoologie* **44**: 79–83.
Stone, A., Sabrosky, C. W., Wirth, W. W., Foote, R. H. & Coulson, J. R. (Eds) 1965. *A Catalog of the Diptera of America North of Mexico*. Agricultural Handbook, United States Department of Agriculture No. 276, iv + 1696 pp.
Strong, L. 1981. Dermestids – an embalmer's dilemma. *Antenna* **5**: 136–139.
Stubbs, A. & Chandler, P. (Eds.) 1979. A Dipterist's Handbook. *Amateur Entomologist* **15**(1978): x + 255 pp.
Sturtevant, A. H. & Wheeler, M. R. 1954. Synopsis of Nearctic Ephydridae (Diptera). *Transactions of the American Entomological Society* **79**(1953): 151–261.
Suenaga, O. 1959. Ecological studies of flies. 5. On the amount of the flies breeding out of several kinds of small dead aimals. *Endemic Diseases Bulletin of Nagasaki University* **1**: 407–413 [in Japanese with English summary].

Suenaga, O. 1963. Ecological studies of flies. 8. The diurnal activities of flies attracted to the fish baited trap. *Endemic Diseases Bulletin of Nagasaki University* **5**: 136–144 [in Japanese with English summary].
Suenaga, O. 1969. Age grouping method by ovariole changes following oviposition in females of *Musca domestica vicina*, and its application to field populations. *Tropical Medicine* **11**: 76–90.
Sullivan, W. N., Du Chanois, F. R. & Hayden, D. L. 1958. Insect survival in jet aircraft. *Journal of Economic Entomology* **51**: 239–241.
Tamarina, N. A. 1958. [Technique of rearing *Calliphora erythrocephala* Mg. in the Laboratory.] *Zoologicheskiĭ Zhurnal* **37**: 946–948 [in Russian with English summary].
Tamarina, N. A. 1967. The study of corpus allatum in living larvae of the blowfly *Calliphora erythrocephala* Mg. (Diptera, Calliphoridae). *Entomologicheskoe Obozrenie* **46**: 283–294.
Tao, S. M. 1927. A comparative study of the early larval stages of some common flies. *American Journal of Hygiene* **7**: 735–761.
Teschner, D. 1961. Zur Dipterenfauna an Kinderkot. *Deutsche Entomologische Zeitschrift* (N.F.) **8**: 63–72.
Teskey, H. J. 1960. A review of the life-history and habits of *Musca autumnalis* De Geer (Diptera: Muscidae). *The Canadian Entomologist* **92**: 360–367.
Teskey, H. J. 1969. On the behaviour and ecology of the face fly *Musca autumnalis* (Diptera: Muscidae). *The Canadian Entomologist* **101**: 561–576.
Teskey, H. J. 1981. Key to families – Larvae. *In* McAlpine, J. F. et al. *Manual of Nearctic Diptera* Vol. 1 [pp. 125–147]. Research Branch Agriculture Canada Monograph No. 27, vi + 674 pp, Hull, Quebec.
Teskey, H. H. & Turnbull, C. 1979. Diptera puparia from pre-historic graves. *The Canadian Entomologist* **111**: 527–528.
Theron, J. G. 1972. Chironomidae (Diptera) causing damage to motor cars. *Journal of the Entomological Society of Southern Africa* **35**: 361.
Thomsen, M. 1934. Fly control in Denmark. *Quarterly Bulletin of the Health Organisation of the League of Nations* **3**: 304–324, 13 pls.
Thomsen, M. & Hammer, O. 1936. The breeding media of some common flies. *Bulletin of Entomological Research* **27**: 559–587.
Thomson, R. C. M. 1937. Observations on the biology and larvae of the Anthomyidae. *Parasitology* **29**: 273–358.
Thorne, B. L. & Kimsey, R. B. 1983. Attraction of Neotropical *Nasutitermes* termites to carrion. *Biotropica* **15**: 295–296.
Tyndale-Biscoe, M. & Kitching, R. L. 1974. Cuticular bands as age criteria in the sheep blowfly *Lucilia cuprina* (Wied.) (Diptera, Calliphoridae). *Bulletin of Entomological Research* **64**: 161–174.
Utsumi, K. 1958. [Studies on arthropods congregate to animal carcasses, with regard to the estimation of postmortem 'interval'.] *Ochanomizu Medical Journal* **7**: 202–223 [in Japanese with English summary].
Utsumi, K., Makajima, M., Mitsuya, T. & Kaneto, K. 1958. [Studies on the insects congregated to the albino rats died of different causes.] *Ochanomizu Medical Journal* **7**: 119–129 [in Japanese with English summary].
Vargas, E. 1977. *Medicina Legal*. 386 pp, University of Costa Rica, San José.
Varley, G. C., Gradwell, G. R. & Hassell, M. P. 1973. *Insect Population Ecology*. viii + 212 pp, Blackwell, Oxford, etc.
Vinagradova, E. B. 1984. [The flesh fly *Calliphora vicina*. A good object for physiological and ecological studies.] *Trudý Zoologicheskogo Instituta Leningrad* **118**: 1–272 [in Russian].
Vinagradova, E. B. & Zinovjeva, K. B. 1972. Maternal induction of larval diapause in the blowfly *Calliphora vicina*. *Journal of Insect Physiology* **18**: 2401–2409.
Vincent, C., Kevan, D. K. McE., Leclercq, M. & Meek, C. L. 1985. A bibliography of forensic entomology. *Journal of Medical Entomology* **22**: 212–219
Vogler, C. H. 1900. Beiträge zur Metamorphose der *Teichomyza fusca*. *Zeitschrift für Entomologie* **5**: 1–6, 17–20, 33–35.
Walker, A. K. & Crosby, T. K. 1979. The Preparation and Curation of Insects. *Information Series DSIR, New Zealand* No 130, 1–54.
Walker, T. J. (Jr) 1957. Ecological studies of the arthropods associated with certain decaying materials in four habitats. *Ecology* **38**: 262–276,
Wardle, R. A. 1921. The protection of meat commodities against blowflies. *Annals of Applied Biology* **8**: 1–9.
Wardle, R. A. 1927. The seasonal frequency of calliphorine blowflies in Great Britain. *Journal of Hygiene* **26**: 441–464.
Wasti, W. S. 1972. A study of the carrion of the common fowl, *Gallus domesticus*, in relation to arthropod succession. *Journal of the Georgia Entomological Society* **7**: 221–229.
Waterhouse, D. F. 1947. The relative importance of live sheep and of carrion as breeding grounds for the Australian sheep blowfly, *Lucilia cuprina*. *Council for Scientific and Industrial Research Bulletin, Melbourne* **217**: 1–31.

Webb, J. P. Jr, Loomis, R. B., Madon, M. B., Bennett, S. G. & Greene, G. E. 1983. The chigger species *Eutrombicula belkini* Gould (Acari: Trombiculidae) as a forensic tool in a homicide investigation in Ventura County, California. *Bulletin of the Society of Vector Ecologists* **8**: 141–146.

West, L. S. 1951. *The House Fly, its Natural History, Medical Importance and Control*. 584 pp, 176 figs, Comstock, New York.

West, L. S. & Peters, O. B. 1973. *An Annotated Bibliography of Musca domestica Linnaeus*. 743 pp, Dawson, London.

Whitfield, F. G. S. 1939. Air transport, insects and disease. *Bulletin of Entomological Research* **30**: 365–442.

Whiting, P. W. 1914. Observations on blowflies; duration of the prepupal stage and colour determination. *Biological Bulletin of the Marine Biological Laboratory, Woods Hole, Mass.* **26**: 184–194.

Wigglesworth, V. B. 1964. *The Life of Insects*. xii + 360 pp, 36 pls, Wiedenfeld & Nicholson, London.

Wigglesworth, V. B. 1984. *Insect Physiology*, 8th edn. x + 191 pp, Chapman & Hall, London.

Wijesundara, D. P. 1957. The life-history and bionomics of *Chrysomyia megacephala* Fab. *Ceylon Journal of Science* (B) **25**: 169–185.

Wilson, E. M. 1982. My week . . . as a forensic scientist. *Biologist* **29**: 13–14.

Wolfe, L. S. 1954. The deposition of the third instar larval cuticle of *Calliphora erythrocephala*. *Quarterly Journal of Microscopical Science* **95**: 49–66.

Wong, H. R. 1972. *Literature Guide to Methods for Rearing Insects and Mites*. Northern Forest Research Centre, Information Report NOR-X-38, 131 pp, Edmonton, Alberta.

Yovanovitch, P. 1888. *Entomologie Appliquée à la Médicine Légale*. 132 pp, Ollier-Henrey, Paris.

Zakharova, N. F. 1966. On the diapause of the Schisobionts, illustrated by example of Sarcophagidae (Diptera). *Proceedings of the 13th International Congress of Entomology* **1**: 579–580.

Zimin, A. 1948. [Key to third instar larva of synanthropic flies of Tadzhikistan.] *Opredeliteli po Faune SSSR* **28**: 1–115 [in Russian].

Zumpt, F. 1952. Flies visiting human faeces and carcasses in Johannesburg, Transvaal. *South African Journal of Clinical Science* **3**: 92–106.

Zumpt, F. 1956a. Calliphorinae. *Die Fliegen der Palaearktischen Region* **11**(64i): 1–140.

Zumpt, F. 1956b. Calliphoridae (Diptera Cyclorrhapha) Part I: Calliphorini and Chrysomyiini. *Exploration du Parc Nationale Albert, Mission G. F. de Witte* **87**: 1–200.

Zumpt, F. 1958. Calliphoridae (Diptera Cyclorrhapha) Part II: Rhiniini. *Exploration du Parc Nationale Albert, Mission G. F. de Witte* **92**: 1–207.

Zumpt, F. 1965. *Myiasis in Man and Animals in the Old World*. 267 pp, pp, 346 figs, Butterworths, London.

Zumpt, F. 1972. Calliphoridae (Diptera Cyclorrhapha) Part IV: Sarcophaginae. *Exploration du Parc Nationale des Virunga, Mission G. F. de Witte* **101**: 1–264.

Zuska, J. & Lastovka, P. 1965. A review of the Czechoslovak species of the family Piophilidae with special reference to their importance to [the] food industry (Diptera, Acalyptrata). *Acta Entomologica Bohemoslovaca* **62**: 141–157.

Index

This index is intended as an integral part of the book and includes the names of insects, other animals and their products, and artefacts (with their synonyms cross referenced), that may be relevant to forensic or carrion work. It is suggested that a read through will familiarise the investigator with almost any insect-related item likely to prove of forensic significance. In a particular case this may perhaps suggest something so far overlooked! It also reveals where work has been done, on what, and by whom.
Page numbers in bold indicate figures

Abbott, C. E., 145,149
Abdomen, 21
Abortifacients, 149
Absorption of body fluids, 16
Acalyptratae, 81
Acari, eating blowfly eggs, 142, 165
  life-history, 53
  on *Cannabis*, 169–72
  on carrion, 16, 18, 22, 23, 27, 142, **163**
  on man, 166
  on vehicles, 167
  symbiosis with beetles, 142, 165
Acarus siro, 162, **163**, 165
Acetic acid, 37
Acid, 37, 77, 156
Aconitum root, 152
Activation analysis, 62
Adventive species, 13
Aeroembolism, 35
Africa, 77, 104, 111, 112, 117, 171
  East, 152
  North, 113, 117, 150, 152
  South, 112, 117, 118, 120
Afrotropical Region, **54**, 81, 105, 120
Age of, adult flies, 44
  eggs, 46
  maggots, 20, 31, 39, 40, 44, **45**, 46–7, **48**
  puparia, 31, 40, 41
  see also case-histories and appropriate taxa
Aglossa, 16, **155**
  caprealis, 150, **151**
  pinguinalis, 150, *151*
Agouti, 161
Agromyzidae, 171
Ahasverus advena, 169, *170*–2
Aircraft, 154, 167, 168, 171
Air currents, 27
Air dried foods, 146
Air passages, 14
Air speed indicators, 154, 168
Akopyan, M. M., 34, 52
Alaska, 107
Albumins, 51
Alcohol, 37, 44, 77
Alcoholism, 50
Aldabra, 117
Aldrich, J. M., 102
Aleochara, 20, 21, 23
  curtula, 145
Alexander, C. P., 75
Algae, 20

Alkaline tissues, 14
Allergic reactions, 154, 168
Alwar, V. S., 101
Alydus eurinus, pilosus, 158, **159**
Alysia manducator, 34, **155**, 156
Ammonia (NH$_3$), 13, 37, 52, 141
Ammoniacal fermentation, 13, 16, 85
Amphibia, 109, 113
  see also toads, frogs
Anatomical preparations, 94, 96
Andaman Sea, 173
Anderson, J. R., 44
Anurans, 22
Aneurism, 92
Anevrina, 77
Annelids, 81, 101
Anobiidae, 27
Anometaxis, 142
Anthicidae, 147
Anthicus, **141**
  hesperi, 147
  hoeferi, 147
Anthobium, **140**, 331
  atrocephalum, 57, 146
  unicolor, 138
Anthomyiidae, 72
Anthrenus, 147
  museorum, 16
  verbasci, **143**
Antibiotic produced by maggots, 113
Ants, 13, 18, 21, 23, 29, 30, 53, 154, **155**, 156, 160, 173
Aphids, 18, 51, 80, 158, 168
Aphodius, **141**, 149
Aphrodisiacs, 149
Apidae, 21, 154–5
Apis mellifera, 154, **155**
Apoplexy, 92
Apparatus, 36–8
Apteromyia claviventris, 87
Aquatic situations, 25–7, 43, 77, 80
Arachnida, 18, 53, **163**, 165, 166
Araneae, 18–20, 22, 53, **163**, 165–6
Archaeological sites, 23, 94
Arctic, 28–9, **54**, 113
  see also Polar Regions
Ardö, P., 79
Argentina, 111
Arnaud, P. H., 99, 169, 172
Arrow poisons, 149, 152
Arsenic acid poisoning, 35
Arson, 52
  see also burnt bodies
Arteriotomy, 35

Arthropods, 53, 156–6
Aruzhonov, A. M., 12, 25, 30, 167
Asellus, 166
Asia, 77
Asilidae, 99
Askew, R. R., 156
Asphalt lakes, 94
Aspirators ('pooters'), 36–7
Assis Fonseca, E. C. M., 98, 125–6, 128, 171
Atheta, 21, 59
  aquatica, 138
  cadaverina, 138
  fungivora, 138
  ravilla, 138
  sordida, 138, **139**
Atomaria ruficornis, **139**
Attacus, 171
Attagenus pellio, 16, **143**, 147
Attractants, 50, 168
  see also baits
Aubertin, D., 109, 112
Augustine mound (Canada), 23
Austen, E. E., 51, 150, 152
Australia, 15, 18, 20, 53, 105, 111–2, 117–8, 120, 158
Australian Region, **54**, 77, 81, 101, 113, 117, 120, 171
Autumn, 19, 46, 53, 104, 107
  see also Seasons
Azarelius sculpticollis, 173
Azelia macquarti, 128

Babies, 13, 27, 49, 122, 154
  see also Child, Foetus, Infanticide
Backlund, H. O., 65
Bacon, 51, 90
Bacteria, 14, 17, 20, 23, 27
Badgers, 36
Bailey, 44
Baits, 36, 50, 73, 84, 94, 142, 147
Balfour-Browne, F., 144
Baranowski, R. M., 9, 158
Bark, 148, **157**, 160
Barnes, J. K., 81
Barns, 150
Basden, E. B., 44
Bass, W. M., 15, 17, 99, 104, 107, 120
Bat excrement, 82
Batra, S. W. T., 173
Baumgartner, D. L., 116
Beaver, R. A., 75, 128
Bed-bug, 158
Bedding, infant, 125
  see also Cot-blankets

Beef, 51, 57, 101
Bees, 21, 53, 154–6, 161, 168
Beetles, 30, 33, 41, 53, 138–49
  see also Coleoptera
Behaviour, 15
Belgium, 101
Bellamy, D., 9, 28–9
Bergeret case, 11, 56
Berlese fluid and mountant, 39
Berndt, K. -P., 61
Beyer, J. C., 35
Bhaskaran, G., 39
Bianchini, G., 14
Bibio, 167
Bibionidae, 167
Biocenoses, 15
Birds, 152, 162,
Birds' nests, 79, 85, 130, 147–9, 152–3
Bishopp, F. C., 112
Black putrefaction, 17, 20
Black, R. J., 63
Bladder, 14
Blair, K. G., 79
Blaniulus guttulatus, 165
Blatella germanica, **155**
Blattodea, 18, 53, 155
Blepharida, 149
Bloated stage, 13, 17–8, 24, 25–7, 80, 158
  in water, 25–7
Blood, 14, 21, 34, 51, 66, 158
Blood-stained clothing, 51, 66, 122, 128, 131
Blood sucking, 158
Blower, J. G., 165
Blowflies, 11, 13–4, 17–8, 20, 23–5, 28, 42, 51, 141
  attractants, 51
  keys to, 40
  see also Calliphoridae
Blowfly Recording Scheme, 51
Bluebottles, 33, 101–3
  see also Calliphoridae
Boats, 168
  see also Ships
Body louse, 162, **163**
Bohart, G. E., 77, 94–6, 99, 101, 118
Bombus, 154
Bone factory, 126
Bones, 27, 90, 146–7, 152, 161
Bone-skipper, **93**, 94, **134**
Borborus ater, **88**, 89
Boreellus atriceps, **28–9**
Borgmeier, T., 80
Bornemissza, G. F., 9, 15, 17–8, 52, 65, 160
Borneo, 173
Boving, A. G., 138, 144
Braack, L. E. O., 52
Braconidae, 21, 34, 154–6
Bradysia, 9, 169–70
Brain, 14, 57
Brash, J. C., 56
Brazil, 117–8
Breweries, 96
Brindle, A., 9, 68, 75, 85, 92
Brine, 77, 94
Brisard, C., 164
Britton, E. B., 149

Broadhead, E. C., 33, 73
Bromdiethylacetyl urea, 35
Brown house-moth, **151**, 152
Bruchidae, 173
Bruchidius mendosus, 173
Bugonia, 80
Bugs, 53
  see also Hemiptera
Bullocks, 101
Bumble bees, 154
Burger, J. F., 11, 15
Buried corpses, 16–25, 49, 56, 59, 75, 79, 84, 129, 146, 148, 156, 165
Burnt bodies, 27, 35, 50–2, 63, 112
Burrow, animal, 152
  see also Nests
Bushmen, 149
Busvine, J. R., 113, 125, 150, 164
Butchers, 171
Butchered meat, 51
  see also Slaughterhouse
Butter, 150
Butterflies, 150, 167
Butyric fermentation, 16–7, 20

Caddis flies, 25, 53, **163** (larva)
Caenis moesta, 168
Calcutta, 117
California, 145, 166
Callahan, P. S., 167
Calliphora, 17, 30, 33–4, 46, 51–5, 69, 70, 103–06, 109, 116, 120, 129, **135**, 146, 156, 165
  keys, 104–5
  alpina, 104–5
  croceipalpis, 103
  loewi, 104–5
  subalpina, 104–5
  uralensis, 20, 104–5
  vicina (*erythrocephala*), 16, 20, 28, 30, 31, 44, 46–9, 56–67, 103–06 132, **135**
  vomitoria, 16, 20, 28, 46, 59, 105
Calliphoridae, 13, 16, 19, 30, 47, 50, 72, 101–3, 141, 156, 161
  egg hatching, 30
  keys, 102–3
  parasites of, 13, 34, 156
  predators of, 13, 142, 156, 165
Callitroga (*Cochliomyia*), 17, 20, 120
Caloglyphus, 23
Calyptratae, 99
Cambala annulata, 23, 165
Camel, 101
Cameroun, 150
Campanotus, 156
Campbell, E., 63
Canada, 23, 28, 92, 105, 107, 113, 120
Canary Isles, 118
Canberra, 117
Cannabis, 12, 77, 152, 169–73
  sativa var *indica*, 173
Canneries, 96
Canteens, 96
Cantharides, 149
Cantharis, **139**
Cantrell, B. K., 102
Cape Verde Islands, 117

Car, 57, 87, 167–8
  accidents, 154
  boot (trunk), 47, 52, 67
  exhaust fumes, 34
  headlamps, 167
  mudguards, 167
  paintwork, 167
  radiators, 50, 167
  tyres, 50, 167
  windscreens, 167
Carabidae, 144, 173
Carabus, **140**
Caravan routes, 89
Carbon dioxide (CO_2), 13, 43
Carbon monoxide (CO), 34
Carbon tetrachloride, 37
Carpet beetles, **143**, 147
Carpets, 60, 67
Carpophilus, **141**, 148
Carrion, attraction to, 50, 142
  fauna, 68–165
  Succession on, 13–35
  see also under animal names, baits
Carter, D. J., 149, 150, 152, 172
Cartilage, 161
Cartín, V. M., 15, 30, 157
Case bearing clothes moth, 152
Caseic fermentation, 16, 85, 98, 122
Caterpillars, 53, 150
Catopid beetles, 84
Catops, **140**, 145,
  picipes, 138–9
  tristis, 138, 145
Cat-flea, 162
Cats, 14, 145, 152, 162
Cattle (see cows), 89, 113, 125
Caustic potash (KOH), 39, 44
Caves, 82
Cecidomyiidae, 39, 171
Cellulose feeders, 161
Cemeteries, 131
Centipedes, 23
Centrophlebomyia anthropophaga, 94
  furcata, 9, 10, **93**, 132, 134
Cephalopharyngeal skeleton, 42–3, **45**, 47
Cercyon, 138–**39**, **140**, 144
  analis, 144
  lateralis, 144
  terminatus, 144
Cereals, 171
Cess pits, 96,107, 118, 122
Ceylon (Sri Lanka), 113, 118
Chaetopodella scutellaris, 87–90
Chafers (beetles), 148
Chalcididae, 173
Chandler, P., 39, 52, 68
Chapman, R. F., 55, 87, 92, 144, 145
Chapman, R. K., 125
Cheese, 51, 90, 150, 171
Cheese-skipper, 90
Cheesy odour, 17
  see also caseic fermentation
Chemical pollution, 25
Chemist's shops, 81
Cheong, W. H., 117

Chicken, 29, 92, 122
  see also fowl, poultry
Child, 56, 57, 146, 166
  see also baby, foetus, infanticide
Chiggers, 166
Chillcott, J. G., 122
Chile, 126, 149
Chilled meat, 50
Chimneys, 27
China, 11, 112–3, 120, 173
Chinery, M., 55
Chironomidae, 155, 167
Chloral hydrate, 39
Chloropidae, 27
Chrysomya, 13, 30, 34, 102, 103, **115**–20, 156
  keys, 115–120
  albiceps, 27, 30, 34, 72, **115**–7, 119
  bezziana, 115, 116, 118
  chloropyga, **115**–7, 120
  inclinata, 116
  latifrons, 118
  mallochi, 116–118
  marginalis, 30, 34, 116, 119, 120
  megacephala, 116–18, 132, **135**
  'nigripes', 94
  putoria, 117
  regalis, 30, 116–20
  rufifacies, 116–117
  varipes, 116, 117
Chrysomelidae, 149
Chu, H. F., 17
Cimex, 158
Cirripedes, 25, 166
Cladocera(beetle), 149
Clambidae, 138
Clark, C. U., 144, 146, 149
Clausen, C. P., 146, 156
Clearing specimens, 39, 44
Cleridae, 16, 19, 21, 146–7
Clethrionomys glareolum, 24, 32
Clogmia albipunctatus, 75
Clothes moths, 27, 150–3
Clothing, 66, 122, 152, 162
Coboldia (Scatopse) fuscipes, 169
Coccidae, **18**
Cochliomya (Callitroga), 102–4, 107, 116
  hominivorax (americana), 120
  macellaria, 25, **119**, 120
Cockburn, A., 92
Cockroaches, 53, **155**, 157
Cocoon, 53, 129, 131, 150, 152, 162
Coe, M., 30, 34, 52, 117, 161
Coe, R. L., 75
Coelopa frigida, 65, 81, **83**
Coelopidae, 65, 71, 73, 81, **83**
Coffin fly, 79, **133**
Coffins, 79, 131
Cogan, B. H., 39, 52
Cold weather, 30
  see also seasons, winter
Cole, F. R., 68
Cole, P., 53
Coleoptera, 13, 18, 22, 23, 138, **139**, **140**, **141**–2, **143**–9
Collecting equipment, **36**, 37, **38**, 39

Collembola, 13, 18, 23, 160, 162, **163**
Colletidae, 21
Colless, D. H., 68
Collin, J. E., 84
Colour, of cars, 167
  of corpse, 13, 17, 25
  of *Lucilia* adults, 109
  of puparia, **39**, 40, 53
Colyer, C. N., 79, 92
Competition, 34, 109, 138, 141, 142, 165
Compost, 77, 122
Conicera, 16, 21, 77–80, **133**
  tibialis, 79
Conistra vaccinii, 150
Containers, insects in commercial, 50, 171–2
  for carcases, 51
  for insects, **36**–43
  plastic, 171–2
Cook, E. F., 169
Cooked meats, 27, 50
Copenhagen, 166
Coproica acutangula, 87–9
  ferruginata, 87–9
  pseudolugubris, 87, **88**, 89
  vagans, 87–9
Copromyza equina, 132, **136**
  similis, 87–9
  stercoraria, 87, **88**, 89
Corbet, A. S., 142, 147, 148, 150, 172
Coreidae, 158
Cornaby, B. W., 9, 22, 30, 90, 96, 156–7
Corylophidae, 149
Corynetes, 16
Costa Rica, 15, 30, 96, 156–7
Cot blankets, 122
Cothran, W. R., 34, 113, 120
Cousin, G, 53
Covered or concealed bodies, 46, 56–8
  see also wrappings
Cow dung, 73, 75, 85
Cows, 82, 101, 152
  see also cattle
'Crab' (pubic louse), 162, **163**
Crabs, 25, 166
Cragg, J. B., 51, 53, 63, 109, 111
Cranium, 161
  see also skull
Crates, 172
Cremation, 24
  see also burnt bodies
Creophilus, 20
  maxillosus, 146
Crickets, 157
Crop, of larva, 44
Crosby, T. K., 39, 172
Crosskey, R. W., 68
Crossley, D. A., 165
Crows, 152
Crowson, R. A., 138, 142, 149
Crustacea, 25, 53, 162, **163**, 166
Cryptophagidae, 23, **141**, 149
Cryptophagus, **141**
Cryptolestes, 172
  ferrugineus, 169, **170**

pusillus, 169
Cryptopleurum, 138, **139**
  minutum, 144
Ctenocephalides canis, 162
  felis, 162
Cucujidae, 169, **170**, 171–2
Cudico (Chile), 149
Culex pipiens quinquefasciatus, 27
Cupboards, 25
Cushing, E. C., 9, 51, 119
Cut surfaces, 14
Cuthbertson, A., 117, 126
Cyclorrhapha, 77
  life-history, 53
Cydnidae, 158
Cynipidae, 24, 154, 156
Cynomya, 16
  mortuorum, 20, 62, 103, **108**, 146
Cynomyopsis, 17, 104, 120
  cadaverina, 25, 46, 103

Dahl, C., 73, 75
Dahl, R., 96
Daily rhythms, 33
Dairies, 171
Darwin, C., 149
Dasyphora, 125
Davies, W. M., 112
Davis, W. T., 112
Dear, J. P., 12, 51, 102, 107
Death, anticipation of, by *Lucilia*, 112
  changes following, 13, 14
  cause of, 34, 49–50
  metabolic state at, 13
  season of, 52
  time of, 13, 17, 56–67
  thermal, points of insects, 30
Debris feeders, 162
Decay curves, 14, **15**, 24
Decomposition, 14, **15**, 23, 24, 51
  by poisoning, 35
  in water, 25, **26**
  stages of, 14–27
Deeming, J. C., 90
Deer, 93
Deflation, 23
Delfinado, M. D., 68
Demerec, M., 98
Demodex, 166
Dendrophilus, 146
Denmark, H. A., 167
Denno, R. F., 34, 113, 120
Deonier, C. C., 30
Dermaptera, 18, 53, **163**
Dermestes, 16, 18, 20, 27, 146, 147, *caninus*, 19
  lardarius, **143**
  maculatus, 16, **143**
Dermestidae, 13, 16, 20–2, 27, 142, **143**, 147
Desiccation, 27, 43, 50, 51
Desmometopa, 99
  m-nigrum, **170**, 171
  tarsalis, 99, 169–171
  varipalpis, 99, 169, **170**, 171
de Stefani, T., 12, 15
Dethier, V. G., 142
Detinova, T. S., 44
Detritus feeders, 162

Development, rates of, 20, 30, **31**, 43, 46, 52, 105, 111, 117–9, 122, 125
arrested, 53
Calliphora, 105
Chrysomya, 117–9
Fannia, 122
Lucilia, 111
Musca, 125
Devon, 66
Dew, 168
Diamphidia nigroornata (*locusta*), 149
Diapause, 53, 63
Diapriidae, 24, 156
Dictyoptera, 155, 157
Diel periodicity, 33
Dienerella filiformis, **170**, 171
Diet, effect on corpse fauna, 50
Dilophus, 167
Diplonevra, 77
Diplopoda, 18, 165
Diptera (flies), 11, 18, 22, 27, 53, 68–137, 158, 169
keys to families, 68, 71–73
parasites of 145, 156
see also separate families and genera
Discomyza incurva, 94, **95**, 96
maculipennis, 94
Disintegration of corpse, 13, 14
Disinterments, 92, 160
see also exhumation
Disney, R. H. L., 80, 81, 152
Dispersal of maggots from carrion, 30, 34, 39, 41, 51, 117, 118
see also migratory phase
Diurnal periodicity, 50
Dog, 15, 19, 20, 59, 77, 150, 156–8
flea, 162
food, 87
*Dohrniphora*1, 21, 77
incisuralis, 23, 80
Donkeys, 94
Donnelly, J., 51
Dor – beetles, 149
Dorsey, C. K., 145
Downes, J. A., 167
Draber-Mońkó, 101–2
Drain pipes, 94
Dried fruit, 148, 171
Drone fly, 80
Drosophila, 9, 98
affinis, 96
ananassae, 96
busckii, 96, **97**, 169, 171
confusa, 96
funebris, 96, **97**, **132**, 134
melanogaster, 96, **97**, 98
phalerata, 96, **97**
quinaria, 96
subobscura, 96, **97**
Drosophilidae, 16, 27, 71, 73, 85, 96, **97**, 98, **132**, 171
Drowned corpses, 25, 35, 50, 142, 162
Drugs, 35, 50
Dry carcasses, 16–18, 20, 27, 30, 141, 149, 152, 157, 165
foods, 90

Dryomyza anilis, 81, **82**
Dryomyzidae, 81, **82**, 132–3
Duda, O., 87
Duffield, J. E., 92
'Dumped' bodies, 46, 52, 167
Dung, 77, 85, 101, 109, 117, 125, 144, 148–9, 165, 171
bibliography of insects on, 90
see also under animal names, excrement, faeces, manure, sewage
Dung-beetles, **141**, 148, 149
Dunn, L. H., 77

Ears, 21, 28, 63, 112
Earwigs (Dermaptera), 18, 53, **163**
Earthworms, 79, 101
Easton, A. M., 11, 56–7, 138
Ecology of carrion, 13–35, 50, 52
Ectoparasites, 13, 25, 162, **163**, 164
Egglishaw, H. J., 81, 90
Eggs, 52
blowfly, hatching, 30, **42**, 104
blowfly, retention in body, 34
broken (hens) *Muscina* in, 130
see also oviposition
Egypt, 101
Egyptian mummies, 27, 92, 117
tombs, 23
Elachisoma aterrima, 87
Electron microscope, 39
Elephant, 30, 117, 161
Ellef Ringnes Island (Canada), 28
Elton, C. S., 52, 138
Emden, F. I. van, 102
Emergence of adult flies, 44, 53
Endomychidae, 149
Endrosis sarcitrella, **151**, 152
Entrails, 14
Ephemeroptera, **155**, 168
Ephestia, 151
Ephydridae, 16, 71, 73, 81, 94, **95**, 96
Erzinclioğlu, Y. Z., 33, 39, 73, 75, 85, 87, 104, 107, 117
Eristalis, 16, 68, 72, **76**, 132, **133**
tenax, **76**, 80, **133**, 154
Erotylidae, 149
Eulalia, 77
Eulophidae, 156
Euparal, 39
Euphyllodromia angustata, 157
Eutrombicula belkini, 166
Evaniidae, 156
Evans, A. C., 53
Evans, G. O., 166
Evaporation of sanious fluids, 16
Evenhuis, N. L., 68
Excrement, 75, 77, 85, 122, 149, 165
human, 85, 90, 96, 99, 101, 107, 125–6
see also under animal names, dung, faeces, manure, sewage, slurry
Exhumed bodies, 17, 131
Experimental work with carrion, 14, 49
Exposed corpses, 13–17
Eyelashes, 166
Eyes, 28, 58, 63, 125, 147, 154

Fabre, J. H., 105, 142
Face-flies, 122, 125
Faeces, 15, 50, 101, 105, 111, 125–6, 128–9
see also under animal names, dung, excrement, manure, sewage, slurry
Fair Isle, 107
False clothes moth, 152
False greenbottles, 125
Fan, C. -T., 102, 120
Fannia, 16, 28, 65, 120–122, 126
key, 122
canicularis, 28, 64, 70, 120, **121**, 122, 132, **136**
manicata, 96, **121**, 122
pusio, 94
scalaris, **121**, 122
Fanniidae, 16, 72, 120, **121**
Farm slurry, 85
Fat body, 44, 73
Fats, 16, 51
Faunal succession, 13–35
Feathers, 152, 162
Fecundity of adult flies, 34
Fermentation, 13, 16–7, 96, 171
Ferrar, P., 120
Ficus microcarpa, 173
Fielder, O. G. H., 112
Fig, 173
Figitidae, 156
Finland, 20, 41, 58–67, 107, 112, 128, 150
Fireplace, 56, 57
Fish, 20, 29, 51, 94, 148, 149
dry, 147
markets, 118
Fishermen, 167
Fishermen's maggots, 107
Fixing larvae, 37
Flat grain beetles, 172
Fleas, 25, 162–3
Fleece, 111
Flesh flies (Sarcophagidae), 33, 99
Flies
see also Diptera
Flight, duration of 142
periods, 33
Floorboards, bodies under, 27, 52, 152
fleas between, 162
Flora (intestinal), 14
Florida, 167
Flotation method of extraction, 44
Flour, 148
Flowers, 147
Fluids, body, 16, 24
Fluorescent dust, marking with, 51
Foetus, 25, 28, 126
see also baby, child, infanticide
Folsom, J. W., 160
Folsomia fimetaria, 23
Food, availability of, 34
Foot, B. A., 84
Footwear, 20, 167
Forced air extraction, 44
Foreman, F. W., 17
Forficula auricularia, 162–3
Formalin, 37

Formic acid, 156
Formicidae, 18–9, 21–2, 30, 173
  in *Cannabis*, 173
Forster, H., 92
Fourman, K., 19
Fowl, 122
  see also chicken
Fox, 29, 34, 59, 81–2, 85, 92, 96, 122, 128
Fraenkel, G., 39, 53
France, 96, 138, 144–5, 160
Fraser, A., 53
Fredeen, F. J. H., 87
Freeman, P., 9, 169, 171
Freezing, effect of, 50–2, 57, 99, 104
Freidberg, A., 9, 94
Fresh corpse, defined, 16, 18
Freude, H., 142
Frog, 29
Fruit, 51, 105, 128, 147
  canneries, 96
  dried, 148, 171
Fruit flies, 25
Fucus, 85
Fuller, M. E., 15, 34, 51, 120, 146–7, 156
Fungi, 14, 23, 27, 33, 77, 81–2, 85, 96, 147, 149–50, 161, 165, 169, 172
  see also *Phallus*, stinkhorn
Furs, 147, 152, 163
  stores, 90
Furnishings (soft), 152

Galleries on carcasses, 161
Gall midges, 39, 171
Gall wasps, 154
Gamasid mites, 18, 20
Gamekeepers' gibbet, 92
Gammerus pulex, 162, **163**, 166
Garbage, 90, 120, 122, 125
Garnett, W. B., 84
Gas generation, 13, 25
Gassing, 34, 50–1
Genitalia of insects, 37
Geographical location, importance of, 28, 49–50
Geometridae, 150
Geophilidae, 18
Geotrupes, **141**
  stercorosus, 58, 149
Geotrupidae, 149
Germany, 19, 80, 84, 156
Ghauri, M. S. K., 158
Gilbert P., 55
Gill, G. D., 84
Gisbert, J. A., 12
Glaister, J., 56
Globulins, 51
Glucose, 39
Glycogen, 13
Glycyphagus, 165
Goat dung, 117
Goddard, W. H., 89–90
Gonzales, T. A., 149, 164
Gonocnemis minutus, 173
Graham-Smith, G. S., 17, 126, 128, 156
Grain, 172

Graphomya maculata, 128
'Grass'
  see *Cannabis*
Grassé, P. -P., 55
Grasshoppers, 157
Graves, 80, 92, 160
Grease, coating on cars, 168
  moth, 150
Green, A. A., 14, 51, 53, 63–4, 99, 103
Green, C. T., 102
Greenberg, B., 44, 101–2, 113, 116, 120
Greenbottles, 33, 108–13, 141
  false, 124
  see also *Lucilia*
Greenfly
  see Aphids
Greenland, 107
Green-staining (of body), 13
Greig, J., 94
Grensted, L. W., 75
Gressitt, J. L., 77, 94–6, 99, 101, 118
Grodowitz, M. M., 39
Groth, U., 61
Ground beetles (Carabidae), 144
Gryllidae, 18
Guam, 99, 101, 118
Guiart, J., 11
Guimarães, J. H., 116–8
Guinea pigs, 20, 158
Gum arabic, 39

Haemoglobin, 34, 51
Hafez, M., 101
Hair, 160
Hair follicle mite, 166
Hairy fungus beetle, 172
Halcitidae, 21
Halidayina spinipennis, 87–9
Hall, D. G., 99, 102, 104, 107, 120
Hall, D. W., 148
Halstead, D. G. H., 146
Halticinae, 149
Ham, 90
Hancox, M., 34
Hamilton, C. J., 55
Hammer, O., 65, 87
Hammond, C. O., 92
Hanski, I., 28, 30, 33–4, 109, 116–8
Hard surfaces and maggot dispersal, 50
Hardy, D. E., 68
Hare (Arctic), 28
Harpalus, 138, **139**
  rufipes, 144
Harris, K. M., 39
Harris, W. V., 161
Hartley, J. C., 80
Häsänen, E., 62, 65
Haskell, P. T., 142
Hawaii, 101
Headless body case, 67
Headlights (car), 167
Head louse, 162, **163**
Heart, 14, 163
Heat produced by maggots, 30
Hebrides, 105
Hecamede albicans, 94–5
  persimilis, 94, **95**

Height of carrion above ground, 52, 142
Helcomyza, 81
Heleomyza, 82
Heleomyzidae, 21, 25, 71, 73, 81–4
Heleotropism, 31, 33
Hemiptera, 53, 158, **159**
Hemp, 152, 173
  see also *Cannabis*
Hendry, W. J., 92
Hennig, W., 12, 52, 68, 85, 93, 95–6, 98–9
Henriksen, K. L., 138, 144
Hepburn, G. A., 17
Hermetia illucens, **76**, 77, 169–70
Herms, W. B., 50
Herniosina bequaerti, 89
Hesperiidae, 150
Heteroptera, 18, 158–9
Hewitt, C. G., 125
Hewson, R., 34
Heymons, R., 145
Hickin, N. E., 9, 162
Hides, 147
Himalayas, 173
Hinton, H. E., 142, 146–8, 150, 152, 172
Histeridae, 16, 18–9, 21–2, 27, **138**, **140**, 146, 149
Hister, 16, **140**
  abbreviatus, 146
  cadaverinus, **138**–9
  striola, 59, 146
  unicolor, 59
Hitch-hiking girls, cases, 58–9, 61
Hobson, R. P., 14
Hoffman, R. L., 165
Hofmannophila pseudosprettella, **151**–2
Holarctic distribution, 175
Holdaway, F. G., 156
Holds of ships, 89, 167
Hollis, D., 55, 68, 96, 102, 142, 146, 157, 161
Holzer, F. J., 25
Honey bee, 154, **155**
Honeydew, 51, 168
Hopkins, G. H. E., 28, 112
Hornets, 25
Horseflies, 68
Horses, 94
Hospitals, 81, 171
Hot springs, 94
Houseflies, 68, 122, 125
  see also *Musca*, *Fannia*
Houses, 130
Hoverflies, 80, 126
Howard, L. O., 90, 99, 125
Howden, A. T., 15, 144, 146, 148–9
Howe, R. W., 148
Hudson, H. F., 112
Hughes, A. M., 166
Hughes, A. W. McKenny, 56
Human corpses, natural stages of decomposition, 13, 14, 17, 25
Humidity, 30
  in life-history, 53
  on site, 41
Hungary, 58
Husky dog, 28

Hydrocyanic potassium
    poisoning, 35
Hydrogen sulphide (H$_2$S), 13
Hydrophilidae, 21, 25, 144
Hydrotaea, 13, 25, 34, 43, 128, **129**
    dentipes, 96, 128, **129**, **131**, 132,
        136
    occulta, 132, 136
Hymenoptera, 13, 18–20, 22, 24,
    53, 154, **155–6**
    in *Cannabis*, 173
    Life-history, 52
Hypnotics, poisoning, 35
Hypogastrura armata, 23
    bengtssoni, 160
Hyponomeuta malinellus, 60

Ichneumonidae, 154–6
Identification of insects, 46, 49,
    53–55
Illingworth, J. F., 145
Immature stages of insects, 52–3
    see also appropriate groups
Imms, A. D., 55, 113
India, 101, 105, 113, 117, 126, 173
Indian meal moth, 171
Indo China, 173
Indonesia, 173
Infanticide, 51, 56–7, 154
    see also Babies, child, foetus
Insecticides, 43, 77
Instars (larval), 52
Intermuscular tissue, 14
Internal organs, 14
Intestinal flora, 14
Intestinal myiasis, 80, 99, 118, 171
Intestines, 14
Ireland, 107, 144
Ischiolepta pusilla, 87, **88–9**
Ishijima, H., 102, 120, 128
Isomegalendiagram, 9, 48, 105
Isotoma sepulchralis, 160
Isopoda, 18
Isoptera, 52, 161
Israel, 94
Itch mite, 166

James, M. T., 82, 102, 113, 120
Japan, 15, 113, 120, 128, 156
Japygidae, 18
Jirón, L. F., 15, 30, 157
Johannesburg, 101
Johnson, M. D., 15
Johnston, W., 15–6, 92
Joy, N. H., 142
Joyce, C., 172

Kalahari Desert, 149
Kamal, A. S., 9, 46–7, 61–4, 101,
    113
Kaneko, K., 9, 78
Kano, R., 12, 102, 109, 112, 120
Karate blow, 59
Kashmir, 94
Kasule, F. K., 146
Kaufmann, R. U., 138
Keh, B., 11, 149
Keilin, D., 43, 75, 128–9, 132
Kelp flies, 81
Kenya, 117, 161

Kerrich, G. J., 55, 68, 96, 142, 146
Kidneys, 14, 50
Killing insect specimens, methods
    of, 36–7
Kilpatrick, J. W., 77
Kimosina (Alimosina) empirica, 87
Kimsey, R. B., 161
King, E. W., 25–6, 77, 90, 96, 145–50
Kishino, M., 9, 30–1
Kitching, R. L., 39, 120
Klasnitzer, B., 142
Klinokinetic mechanism, 142
Knifing, 58, 61, 66, 113
Knipling, E. B., 30, 167
Knipling, E. F., 101–2
Knutson, L. V., 90
Kolb, H. H., 34
Kükenthal, W., 55
Kumaon Himalayas, 173
Kumar, R., 90
Kurahashi, H., 102
Kuusela, S., 34

Laake, E. W., 9, 119–20
Laboratories, insects occurring in,
    171
Laboratory procedure, 41
Labrador, 107
Lactic acid, 13
Lakes, 173
    see also reservoir
Laminaria, 85
Lamp shades, 120
Lane, P., 149
Lane, R. P., 52, 87, 152
Laos, 173
Large tabby moth, 150
Larval stage, anatomy of, 42
    (blowflies), 45
Lastovka, P., 92
Lathridiidae, 49, 169–71
Latrines, 94, 107, 118, 122, 128, 171
Latrine fly, 122
Laurel leaves, 152
Laurence, B. R., 73, 75, 89
Laurikainen, E., 65
Lauxaniidae, 71
Lavatories, 77, 122, 126, 130, 171
Leaf-cutter bees, 154, **155**, 168
Leaf litter, 160, 169
Leaf-miner, 171
Leather, 152
Leclercq, M., 4, 9, 11–3, 17, 30, 49,
    57, 92, 101, 151, 154, 156, 166
Lehrer, A. Z., 102
Leiodidae, **138**, **140**, 145
Lemming, 9, 28–9
Lepidoptera, 16, 18–9, 21–2, 60,
    150, **151–3**, 155
    in *Cannabis*, 169, 171
Lepismidae, 18
Leptocera, 23
    cadaverina, 87
    caenosa, 87, **88–9**
    fontinalis, 87
    pectinifera, 87
Leptometopa coquilletti, 99
    latipes, 98–9
Lesser fruit flies, 96, **97**, 98, **134**
    see also Drosophilidae

Lesser house fly, 120
Leston, D., 158
Levinson, Z. H., 125
Lewis, T., 33
Lice (Phthiraptera), 162, **163–4**
Lichens, 161
Life-histories, of blowflies, 47
    (table)
    of insects, 52
Ligaments, 161
Light, importance of, 17, 31, 53,
    89, 114, 167
    ultra violet, 168
Lightning, 28, 112
Lime (chemical), 94
Limosina silvatica, 89
Lindquist, A. W., 14
Lindroth, C. H., 144
Linnaeus, C., 11
Linssen, E. G., 142
Lithobiidae, 18
Liver, 14, 27, 43, 51, 94
Lizard, 30, 96, 156, 157
Lloyd, J. E., 90
Lobanov, A. M., 84
Lobodromia, 157
Lodge, O. C., 51
Logs, 85, 157
Lopatenok, A. A., 12
Lopes, H. de Souza, 101–2
López, L., 12
Lord, W. D., 11, 15
Lothe, F., 118
Lovebugs, 167
Lowne, B. T., 105
Lucilia, 16–7, 34, 37, 46, 50–1, 55,
    64, 99, 102–3, 107, 108–13, 116,
    120, 129, 146, 156
    keys, 108–113
    ampullacea, 103, 109, **110–112**
    bufonivora, 109, 113
    caeruleiviridis, 25
    caesar, 20, 28, 47 (rate of
        development, table), 58, 109, 112
    cuprina, 28, 113
    elongata, 113
    illustris, 20, 59, 62, 109, **110**, 112
    papuensis, 113
    porphyrina, 113
    richardsi, 62, 109, 113
    sericata, 20, 46, 62–3, 109, **110**,
        111, 132, **135**
    silvarum, 20, 62, 109–113
    anticipating death, 112
    Muscidae confused with, 125
Lundt, H., 9, 19, 21, 52, 80, 84, 156
Lungs, 14
Luther, H., 150
Lycaenidae, 150
Lydney case, 56
Lygaeidae, 158–9
Lyneborg, L., 122
Lytta vesicatoria, 149

McAlpine, D. K., 68
McAlpine, J. F., 9, 28, 68, 92–3, 132
McClung case, 57
McIntosh, R. P., 9
Mackerras, I. M., 55
Mackerras, M. J., 117

McKnight, B. E., 11
MacLeod, J., 51, 52
Macquart, M., 11
Macrocheles muscadomesticae, 169, 172
Madagascar, 111–2, 117–8
Madiza glabra, 16, 99, 132–3, **134**
Madras, 101
Maggots, 11, 25, 42, 68
  heat produced by, 30
  in case-histories, 56–67
  in vehicles, 167
  life-history, 53
Magy, H. I., 63
Maize, 150
Malaysia, 113, 117, 173
Malloch, J. R., 68
Mammals, small, 15, 82, 162
  see also nests
Mangan, R. L., 85
Mant, K., 56
Mantelpiece, 56
Manure heaps, 125, 171
Marchenko, M. I., 12
Marijuana, 169–173
Marine situations, 25, 65, 81, 94, 101, 166–7
Markets (outdoor), 118, 120
Marking insects, 52
Marrow (bone), 24, 161
Marshall Islands, 101
Marshall, J. E., 142
Marshes, 150
Martin, J. E. H., 39
Mason, W. R. M., 154, 156
Massachusetts, 103
Matile, L., 126
Mauritius, 105, 112, 117–8
Mayfly, **155**, 168
Meal, 148
Meal moth, **151**, 171
Mealworms, 143, 148
Mearns, A. G., 56
Meat, 27, 84, 92, 101, 105, 107, 125
  factories, 92
  markets, 118
Medieval excavations, 94
  see also archaeological sites
Meek, C. L., 12
Megalotomus quinquespinosus, 158, **159**
Megaselia, 77, **78**
  rufipes, **78**, 171
Megasternum, 138, **139**
  obscurum, 144
Mégnin, J. -P., 4, 11, 15, 17, 25, 79, 92, 126, 131, 146–8, 151, 165–6
Melandryidae, 149
Melanolestes picipes abdominals, 158, **159**
Mellanby, K., 53, 166
Meloidae, 149
Meloponinae, 161
Mengyu, Z., 102
Mercury content of flies, 62
Meroplius minutus (stercorarius), 85, **86**, 87
Metabolic state at death, 13
Metamorphosis of insects, 30, 53
Meteorological data, 30, 41

in case histories, 56–67
Metopina, 23, 77–80
  subarcuata, 23, 80
Metropolitan Police Forensic Science Laboratory, 10, 154
Mexico, 120
Mice, 60
Michelsen, V., 94
Microchrysa polita, 77
Microclimate, 17, 30, 65
Microgramme arga, 169, 172
Microorganisms, 13
Micropezidae, 161
Microscopical examination, 39, 44
Midges, 155
Migratory phase (dispersal of maggots), 30, 34, 39, 41, 50, 90, 104 117–8, 145
  in case histories, 56–67
Miháyli, F., 58, 90
'Mikrokaverna', 79
Milichiidae, 71, 99, 169, **170**, 171
Milk bottles, 77, 96
Millipedes, 23, **163**–5
Miridae, 158
Mites, 18, 23, 27, 53, 142, **163**, 165–6
  in *Cannabis*, 169
  on vehicles, 167
Mole, 79
Mollusca, 75, 81, 94, 95
  see also Snails
Monopis rusticella, 16, **151**–2
Monotoma 148
Moore, B. P., 144
Moorehouse, D. E., 39, 104, 118, 120
Morley, C., 138
Morpholeria kerteszi, 21, 84
Mortuary, 43
Mosquitoes, 71
  larvae, 27
Moth-flies (Psychodidae), 75, 132, **133**
Moths, 18, 150–3
  see also Lepidoptera
Motter, M. G., 17, 92, 160,
Mould, 17, 172
Moulting of larvae, 44–5, 53
Mounds, termite, 161
Mounting specimens, 39
Mouth, 21, 28, 63
Mucus seeking flies, 125
Mud, as habitat, 144
  on shoes, 50
  -guards on cars, 167
Müller, B., 46, 156
Mummies, 27, 92, 117
Mummification, 27, 29, 52
Musca, 72
  autumnalis, 16, 124–5, 132, **137**, 146
  domestica, 16, 27, 31, 40, 46, 77, 120, 123, 125–6, 132, **137**, 171
  puparium, 131
  rates of development, table, 47
  wing, 70
Muscidae, 13, 16, 19, 72–3, 122–37, 161
Muscina, 9, 21, 65, 128–32
  assimilis, 25, 59, 62

pabulorum, 129, 131 (cocoon & puparium), 132
  stabulans, 16, 66, 70 (wing), 128–9, 132, 133, **137**
Muscle, 13–4, 51 (plasma), 94
Museum beetle, **143**, 147
Musk ox, 28–9
Mutillidae, 156
Mutton, 51
Mycetophagidae, 149, 169, 172, **402**
Myiasis, 11, 75, 80, 99, 113, 117–8, 122, 167, 171
Myodocha serripes, 158, **159**

Nabaglo, L., 9, 24, 32, 52, 90, 144–6
Nagasawa, S., 9, 30–1
Nandi, B. C., 102
Napkins, 122
Nasutitermes nigriceps, 161
National fish skin week, 51
Nearctic Region, 54 (map)
Necrobia, 16, **141**, 146
  rufipes, 146
Necrobia litoralis, 145
  surinamensis, 27, 145
Necrophagous species, role of, 13
Necrophila americana, 145
Necrophorus see *Nicrophorus*
Nectar, 51
Nectar feeders, 150
Nematocera, 73
Nematodes, 17, 20
Nemobius, 157
Nemopoda, 71
  nitidula, 85, **86**, 87
Neoblattella fraterna, 157
Neoleria inscripta, 82, **84**
Neotropical Region, 54 (map)
Neottiophilidae, 92
Nepal, 113
Nesomylacris sp., 157
Nests, ants', 160
  bees', 168
  birds', 79, 85, 130, 147–9, 152–3
  mammals', 82, 147, 149, 152
  rodents', 147
  termites', 161, 173
Nets, for collecting, 36–8
New Caldeonia, 173
New England (U.S.A.), 15
New Guinea, 118,
New Jersey, 142
New Zealand, 105, 172–3
Nicrophorus, 16, 138, **139–40**, 142, 145, 165
Nielsen, B. O., 96, 104
Nielsen, S. A., 30, 104
Niquirana, 157
Nitidulidae, 19, 21, **141**, 147–8
Nitrogen, (N_2), 13
Nixon, G. E. J., 9, 184
Noctuidae, 150
Nocturnal activity of blowflies, 46, 58, 103
  of Sphaeroceridae, 90
Norn, M. S., 166
Norris, K. R., 33, 39, 50, 102, 109, 117
North Carolina, 15

Nuorteva, P., 4, 9, 11, 15, 19–21, 28–30, 33, 43, 49–52, 58–67, 81, 92, 96, 103, 107, 112, 114, 122, 128–9, 131–2, 146, 149
Nuts, 171
Nyctobora noctivaga, 157
Nymphal stage, 53
Nymphalidae, 150

Occupation, effects on corpse, 50
Odontotermes zambesiensis, 161
Odour of decay, 16–7, 25, 27, 58, 63, 82, 120, 142, 171
Oecophoridae, **151**, 152
Oecothea fenestralis, 84
Offal, 126, 128
O'Flynn, M. A., 39, 52, 104, 118, 120
Ohtaki, T., 53
Oiceoptoma noveboracensis, 142
Oil, survival time of *Demodex* in, 166
Okada, T., 9, 96, 98, 171
Okely, E. F., 89–90
Oldroyd, M., 39, 55, 68, 81, 92–4 99, 116
Omnivorous species, role of, 13
Onchiurus, 160
Oncocephalus geniculatus, 158, **159**
Onthophagus, **141**, 149
Opalimosina collini, 87–9
   denticulata, 87–9
Operating theatres, 171
Ophyra, 13, 16, 27–8, 34, 43, 46, 52, 66–7, 126–8, 146
   Key, 126–8
   anthrax, 126–8
   capensis, 126–8, 131
   chalcogaster, 126–8
   leucostoma, 126–8, 132, **136**
   nigra, 126–8
Oriental Region, **54** (map)
Orifices (natural), 52
Orthellia, 125
Orthoptera, 22, 157
Orygma luctuosum, 81, 85, **86**, 87
Oryzaephilus mercator, 172
   surinamensis, 169, **170**, 171, 172
Osten Sacken, C. R., 80
Otitidae, 94
Ovaries (of flies), 44
Oviposition, 27, 30, 34, 46, 50–2, 92
   Calliphora, 103
   communal, 120
   Lucilia, 111
   on vehicles, 168
Owl pellets, 152
Oxen born bee, 80
Ox heart, 73
Oxypoda, 33, **140**
   lividipennis, 146
Oxytelus, 21
   insignitus, 23
Oxygen, 53, 79
Ozanovskii, V. S., 12
Ozerov, A. L., 94

Pacific Islands, 101, 111, 118, 171
Packing materials, 152, 172 (crates)
Paintwork, damage to, 167
Pakistan, 119
Palaearctic Region, **54** (map)
Pampel's fluid, 37
Panama, 77, 120
Panimerus notabilis, 75
Papavero, M., 68
Papilionidae, 150
Papp, L., 89
Paralucilia wheeleri, 120
Paramonov, S. J., 93
Parapristina verticellata, 173
Parasites, role of, 13, 17–8, 20, 34, 44, 52, 65, 102, 145
Parasitic Hymenoptera, 34, 56
Parathion poisoning, 35
Parcoblatta, 157
Parish, H. E., 9, 51, 119
Parmenter, L., 82, 87
Parthenogenetic species, 89
Pasture, 19
Patton, W. S., 117
Pauropoda, 18
Pavillard, E. R., 113
Payne, J. A., 14–5, 17, 19, 21, 23, 25–6, 29, 33, 51–2, 75, 77, 80, 87, 90, 96, 99, 104, 122, 128, 144–50, 154, 156–7, 160, 165
Peacock, E. R., 148
Pediculus humanus v. *capitis*, *h.* v. *corporis*, 162 **163**
Pelecinidae, 156
Perfumers' shops, 81
Periodicity (seasonal & daily), 33
Perth (Australia), 18, 20
Pessoa, S. P., 149
Peters, O. B., 125
Phaenicia, 108
   see also *Lucilia*
Phalangida, 19, 21
Phallus impudicus (stinkhorn fungus), 33, 80, 82, 96, 150
Phaonia erratica, 128
Phalaenidae (= Noctuidae), 150
Pheasant, 81
Pheidole, 156
Pheidologeton diversus, 173
Phenobarbitol, 35
Philippines, 113
Philonthus, 16, 21, 59, 138, **139**
Philosepedon humeralis, 75
Phora aterrima, *vitripennis*, 79
Phoridae, 16, 19, 21, 23, 27, 57, 71, 77–80, 94, 156, 171
Phormia, 53, 156
   regina, 23, 46, 103, 103, 113–4
   terraenovae, 25, 47, 58, 60–1, 103, 113, **114**, 132, **135**
Phosphorus poisoning, 35
Photographs, 44
Phototropism, 31
Phthiraptera (lice), 25, 52, 162, **163**, 164
Phthirus pubis, 162, **163**
Phytomyza horticola, 171
Pickling factories, 96
Pigmentation, of adults, 79
   of larval mouthparts, 46
   of puparia, 39, 41, 43
   of wings, 53

Pigs, 15, 19, 21, 25–7, 29, 51, 77, 85, 96, 104, 144–9, 154, 156, 158–60, 165
pens, 94
sties, 122, 162
Pinning insect specimens, 37
Piophila, 92–3
   bipunctatus, **91–3**
   casei, 16, 27, **91–3**, 132, 134
   foveolata, **91–3**
   nigriceps, 46 (rate of development, table), 91–3
   varipes, **91–3**
   vulgaris, **91–3**
Piophilidae, 16, 18–20, 72, 85, 90–4
Pirates, 58
Pitkin, B. R., 87
Plachter, H., 89
Plant bugs, 158–9
Plants, 11, 28–9, 51, 56–67 (case histories), 51, 120 (with carrion odour), 149, 152 (arrow poisons), 150, 152 (poisons)
Plasma, 51
'Plasticine', 44
Plastic trunk, 171
Plastozote, 44
Plecia nearctica, 167
Plectochetos, 18
Plodia interpunctella, 150, 169, 171
Poecilochirus, 142
Poisoning, 35, 50–1, 149, 152
Poisons, 152
Poland, 23, 144–6
Polar Regions, 28, 113
   see also Arctic, Antarctic
Polietes albolineata, 128
Pollinators, 11, 173
Pollution, 25, 62
Polyclada (beetle), 149
Polyethylene sheet, 58–9
Polythene bags, 34, 46, 56, 67
Polyxenidae, 18
Pompilidae, 154
Pont, A. C., 85, 126
'Pooters', 36–7
Portschinsky, I. A., 131
Post mortem examination, 43
   interval – see time of death
'Pot' – see *Cannabis*
Potatoes, 82
Poultry dung, 126
   farms, 92, 122, 130
Povolný, D., 112
Power stations, 25
Pradhan's formula, 30
Prawns, 25, 53, 166
Predators, role of, 13, 17, 20, 30, 34, 51, 154
Prehistoric graves, 4, 23, 132
Prenolepis imparis, 21, 156
Prepupal stage, 39, 52, 56–67 (case histories)
Preservation, 37–46
Prins, A. J., 118–20
Privies, 122
Proctotrupidae, 21, 156
Protein fermentation, 99
Proteins, 27, 51, 90

INDEX

Proteinus brachypterus, 138
Protophormia terraenovae, 25, 58, 60–1, 103, 113, **114**, 132, **135**
Protozoa, 17
Protura, 18
Protothyreophora grunini, 94
Pseudexia prima, 94
Pseudoscorpions, 18
Psocoptera, 18
Psychoda alternata, 75, 132, **133**
 surcoufi, 75
Psychodidae, 23, 27, 68, 72, 75
Ptecticus trivittatus, 77
Pteromalidae, 156
Pterostichus niger, 144
Ptesimis fenestralis, 89
Ptiliidae, 138, 149
Ptilinum, 41, 52–3
Ptinidae, 27, 142, **143**, 148
Ptinus brunneus, 16
 tectus, 143
Ptomaphila, 18, 20
Pubic hair, 160
 louse, 162
Public houses, 96
Pulex irritans, 162, **163**
Pullimosina heteroneura, pullula, 87–9
Pupal stage, 53, 156
Puparia, 39, 43–4, 53, 56–67 (case histories), **131**
Pupariation, 39, 52
Pupation, 52
 premature, 156
Putman, R. J., 12, 34, 52, 101
Putrefaction, 13–4, 20, 28–9
Putrid liquids, 96, 99
Pyralidae, 16, 150–1
Pyrellia, 125

Quebec, 107

Rabbit, 29, 82, 84–5, 92, 144–5, 152, 156
Radiators (car), 50, 167
Ragge, D. R., 157
Rain, 30, 51, 89, 99, 120, 160, 167
Rancid fat, 16
Rat, 15, 29, 35, 156
Ratcliffe, F. N., 112
Rate of decay, 17, 35
Rate of development (Diptera), 43
 Calliphora, 105
 Chrysomya, 117–9
 Fannia, 122
 Lucilia,
 Musca, 125
 Ophyra, 126
 graph, **31**
 table, 20, 47
Rat-tailed maggot, 80
Réamur, R. A. F. de, 11
Rearing specimens, 36, 41–3, 56–7 (case histories)
Rectal myiasis, 80
Redi, F., 11
Red tailed flesh fly, 101
Reduviidae, 99, 158–9
Reed, H. B., 9, 15, 17, 19–20, 77, 144, 149–50, 157

Refrigerator, 63
Refuse, 125
Regions, zoogeographic, 54 (map)
Reindeer, 113
Reiter, C., 9, 11, 30–1, 48, 105
Remmert, H., 65
Reptiles, 22
Reservoir, 168
Restaurants, 96
Réunion, 112, 117–8
Rhizophagidae, 16, 21, 148
Rhizophagus, 138, **139**, **140**
 parallelocollis, 16, 79–80, 148
Richardiidae, 161
Richards, O. W., 87, 90, 154
Rigor mortis, 13
Ritual suicide, 52
'Roaches (cockroaches), 52, **155**, 157
Roback, S. S., 102
Robber flies (Asilidae), 99
Roberts, R. A., 52
Robinson, G. S., 150, 152
Roch, M., 154
Rodent burrows, 82
 carcases, 15, 23, 144–5
 nests, 147
Rodriguez (island), 117–8
Rodriguez, W. C., 15, 17, 99, 104, 107, 120
Roháček, J., 89–90
Rohdendorf, B., 102
Rohe, D. L., 63
Rondani, C., 92
Rorqual, 126
Rove beetles (Staphylinidae), 138, **139**, **140**
Rowan tree, 60
Roy, D. N., 117
Rozkosný, R., 77
Rozsypal, J., 112
Rubbish tips, 85
Rubezhanskiv, A. F., 12
Russia, 94
Ruxton case, 56

Sabrosky, C. W., 99, 171
St. Helena, 111, 117
St. Kilda, 107
Salt marshes, 94
Saltella sphondylii, 85
Sand, 94
San Francisco, 168
Sanjean, J., 102
Sankey, J. H. P., 87, 92, 144–5
Sap, 147
Saponification of fat, 92
Saprinus, 16, 20, **140**
 aeneus, 146
 planiosculus (= *cuspidatus*), 146
 semistriatus, 146
Sarcophaga, 16–7, 33–4, 37, 99, 101, 103, 107, 120
 argyrostoma, 101
 bullata, 46, 101
 carnaria, 47, 56, 101, 132, **135**
 cooleyi, 46, 101
 crassipalpis, 101
 dux, 101
 exuberans, 101

gressitti, 101
haemorrhoidalis, *100*, 101
hirtipes, 101
inaequalis, 101
inzi, 101
misera, 101
munroi, 101
shermani, 46, 101
tibialis, 101
Sarcophagidae, 16, 19, 30, 47, 53, 72–3, 99, **135**
Sarcoptes scabei, 166
Sargus, 77
Satchell, G. H., 75
Sato, K., 109, 112
Satyridae, 150
Sauna, 67
Saurians, 22
Sausages, 51
Sawflies (Symphyta), 154
Saw toothed grain beetle, **170**, 172
Scanning electron microscope (SEM), 39
Scaptomyza graminum, 96
Scatopsidae, 75, **76**, 169, **170**
Scatopse fuscipes, 172
Scarabaeidae, 21–2, **141**, 148–9, 161
Scathophaga stercoraria, 132, **136**
Scathophagidae, 72, 84, **136**
Schlinger, E. I., 68
Schmitz, H., 17
Schoof, H. F., 77
Schumann, H., 39, 89–90, 108–9
Sciara militaris, 132
Sciaridae, 24, 68, 75, 169, **170**
Scoliocentra villosa, 84
Scorpions, 166
Scotland, 107
Screw worm, 120
Scuttle flies, 77, **133**
Sea, 65
 shore, 50, 65, 85, 94, 101, 145
Seal, 15, 87
Season, 17, 19, 30, 33, 46, 56–67 (case histories); 177, 161 (dry)
Seaweed fly, 65, 81, 90
Sebum, 160
Secondary myiasis in sheep, 117–8
Secondary screw worm, 120
Sedatives, 64
Sedge thatch, 150
Seed feeders, 171
Séguy, E., 9, 112, 130, 132
Senior-White, R. A., 102
Sepsidae, 16, 19, 71, 73, 81, 84–7
Sepsis, 85, (key)
 biflexuosa, 85–6
 cynipsea, 85–6, 133–4
 punctum, 85–6, **87**
 thoracica, 85–6
 violaceum, **87**
Septic tanks, 171
Seshiah, S., 101
Sewage, 25, 85, 90, 94, 122, 171
 disposal tanks, 87
 filter beds, 75
 works, 75, 85
Sewers, 94
Seychelles, 117
Shade, 17, 31, 33, 51, 103

Shaw, M. W., 92
Sheals, J. G., 166
Sheep, 15, 34, 51, 75, 109, 112–3, 117, 128, 147, 152
　strike, 111
Shellfish, 51
Shetland, 107
Shinonaga, S., 102
Shiny surfaces attracting insects, 168
Ships, 167–8, 171
Shoebox, 49, 56
Shoes, 167,
　insects on, 49
Shores, 66, 81, 94, 101
Shorrocks, B., 96, 98, 171
Shrews, 81, 152
Shrimps, 25, 53
Shubeck, P. P., 142
Sickle murder, 11
Siddons, L. B., 117
Siikainen (Finland), 41
Silken cocoons, galleries, 150, 152
Silkmoth larvae, 132
Silkworm, 171
Silpha 16, 138, **139**, **140**, 145
　americana, 145
　noveboracensis, 142
　obscura, 145
Silphidae, 13, 16, 19, 21, 27, **139**, **140**, 142, 144
Silvanidae, 169, **170**, 171
Simmonds, M. S. J., 9, 156
Simmons, P., 90
Simpson, K., 9, 13, 25, 56, 149, 162
Sims, R. W., 166
Sinea diadema, 158, **159**
Sinews, 147
Singapore, 173
Siphonaptera (fleas), 25, 162, **163**
Size
　of adult insects, 34
　of adult *Lucilia*, 109
　of corpse, 34, 113, 120
　of larvae, 34, 46, 105
　of pupae, 34
Skeleton, 73, 161, 165
Skeletonization, 23
Skidmore, P., 84, 125, 128
Skin, 21, 146, 156
　human, 25, 156
　insect, 52
Skins, 90, 146–7, 152
Skirting boards, 152
Skull, 92, 161, 165
Slater, J. A., 9, 159
Slaughter–houses, 14, 92, 113, 118, 120, 126
Sloth, 161
Slurry (farm), 85
Small mammals, 15
Small tabby moth, 150
Smart, J., 90
Smit, B., 112, 117–8
Smit, F. G. A. M., 162
Smith, H. F., 156
Smith, K. G. V., 11–2, 16, 33, 39, 51–2, 55–7, 68, 75, 77, 81–2, 85, 87, 92, 96, 99, 102, 114, 122, 138, 144–6, 149, 160, 162, 166

Smith, W. F., 53
Smoked foods, 146
Snails, 75, 77, 96, 128
　see also Molluscs
Snakes, 29, 166
Snow, 92, 94, 128
Socotra, 117, 119
Soil fauna, 13, 17, 79, 160, 165
　see also Buried corpses
Soldier flies, 77, **170**
Somme, 51
Soós, Á., 68
Sorbus acuparia, 60
South Africa, 101, 105, 117–8
South America, 101
South Carolina, 22, 29, 154, 157, 165
Southwood, T. R. E., 52, 159
Spanish fly, 149
Spector, W., 149
Spelobia cambrica, 87–9
　clunipes, 87, **89**
　luteilabris, 87–9
　palmata, 87–9
　parapusio, 87–9
Sphaerocera curvipes, 87–9
Sphaeroceridae, 16, 19, 23, 27, 71, 73, 87, **88**, 90, **136**, 156
Sphecidae, 154
Sphingidae, 150
Spider beetles, **143**, 148
Spiders (Araneae), 13, 18–20, 22, 23, 53, 99, 154, **163**, 165–7
Spilsbury, B. H., 79
Spiniphora bergenstammi, 77
Spiracles, 43, 53
Spontaneous generation, 11
Spring, 19, 33, 53, 104
　see also Seasons
Springett, B. P., 142, 165
Springtails (Collembola), 13, 23, 53, 160, **163**
Sri Lanka (Ceylon), 113, 118
Stab wounds
　see knifing
Stable-flies, 113, 122, 130
Stables, 130, 150
Stakelberg, A. A., 10, 132, 145–6
Staphylinidae, 13, 16, 18–22, 27, 59, 138, **139**, **140**, 149, 173
Staphylinus maxillosus, 27
Starchy foods, 172
Steele, B. F., 145
Steffan, W. A., 68
Stenus basicornis, 173
Steyskal, G. C., 90
Stings, 154, 168
Stinkhorn fungus (*Phallus impudicus*), 33, 81–2, 86, 150
Stomach, 14
Stone, A., 68
Stones, 157
Stored products, 27, 52, 90, 142, 148, 150, 172
Stove, 57
Stratiomyidae, 68, 71–2, **76**, 77, 169
Straw, 150
Streams, 173
Strophanthus, 152
Stubbs, A., 39, 52, 68

Stunted larvae, 105
Sturtevant, A. H., 96
Submerged corpses, 257
Succession, 5, 13–35
Suenaga, P., 29, 33–4, 44, 99
Sugars, 51
Suicide, 57, 64, 149
Sullivan, W. N., 30, 167
Sumatra, 173
Sumbawa, 173
Summer, see 19, 33
　see also Seasons
Sun effect of, 29, 33, 50–1, 56–67
　(case histories), 109
Suspended animation, 53
Swamp, 96
Swarms of flies, 79
　indoors, 99
　under trees, 118, 120
Sweat-flies, 122
Sweden, 79, 94
Sweet substances, 118, 125
Symbiosis, beetles with mites, 142
Symphyla, 18
Symphyta, 154
Syrphidae, 16, 68, 72, 80, 126, 154
Syrphus, 80
Szyska, M. L., 102

Tabby moths, 150
Tachys, 173
Tamarina, N. A., 53
Tams, W. H. T., 150
Tanneries, 90
Tanning of puparium, 53
Tao, S. M., 108
Tate, P., 75
Taylor, L. R., 33
Taylor, M. E., 87
Techniques, 37–41
Teichomyza, 73
　fusca, 16, 94, **95**
Telmatoscopus, 75
Telomerina pseudoleucoptera, 89
Temperate regions, 29, 157
Temperature, 13, 17, 27, 30, 31
　(effects, graphs), 41 (on site), 43 (in mortuary), 48, 53 (in life-history), 56–67 (case histories), 79, 105, 120, 126
Tenasserim, 173
Tenebrio molitor, **143**, 148
　obscurus, 16, 148
Tenebrionidae, 16, 27, 142, **143**, 148, 173
Tenebroides, **141**
Tennessee (U.S.A.), 15, 19–20, 77, 150, 157
Tephritidae, 25
Termites (Isoptera), 30, 53, 161, 173
Terrilimosina racovitzai, 89
　schmitzi, 89
Teschner, D., 90, 122
Teskey, H. J., 23, 68, 108, 132
Thailand, 173
Thatch, 150
Themira leachii, 85
　nigricornis, 85
　putris, 85, **86**, 87

Thermal death points, 30
Theron, J. G., 167
Thigh (human), 92
Thomas, V., 75
Thompson, F. C., 167
Thomsen, E., 10
Thomsen, M., 10, 39–40, 65
Thomson, R. C. M., 132
Thoracochaeta zosterae, 88, **89**, 90
Thorne, B. L., 161
Three toed sloth, 161
Thyreophora cinophila, 94
Thyreophoridae, 9, 16, 71, 92–3
  see also Piophilidae
Thyrididae, 150
Thysanoptera, 18
Timber, 161
  see also Wood
Time of arrival on corpse, 33
Time of day, 33, 50
Time of death (post mortem interval), 11, 13, 17, 56–67 (case histories)
Tinant-Dubois, J., 30, 156
Tineidae, 16, 18, 20, 147, **151**–2, 172 (in *Cannabis*)
Tineola biselliella, 16, **151**–2
  pelionella, 16, **151**–2, 172 (in *Cannabis*)
Tingidae, 158
Tinned meat, 51, 84
Toad, 30, 96, 113, 156–7
Toilets, 122
Tombs, 24
Topographical location, 30
Traffic accidents, 154
Transport, 46, 167
Transported bodies, 46, 52, 167
Traps, 36, 52, 94
Trichlorethylene, 81
Trichocera, 132, **133**
  annulata, 73–5
  garretti, 9
  hiemalis, 73–5
  maculipennis, 73–5
  regelationis, 73–5
  saltator, 73–5
Trichoceridae, 33, 68, 72, 73–5, **133**
Trichoptera, 25, 53
Trichopterygidae, 18, 20
Trickling filter fly, 75
Tripe, 51
Triphleba, 77
Tristan da Cunha, 111
Trogidae, 19, 21, **141**, 149
Trogossittidae, **141**
Tropics, 22, 29–30, 33, 35, 99, 120, 160
Tropimeris monodon, 173
Trox, **141**, 149
Trunk, bodies in, 46, 51, 171
Tuberculosis, 149
Tullbergia, 160
Tunnels beneath carcases, 148–9
Turnbull, C., 23, 132
Turtle, 161
Tyndale-Biscoe, M., 44
Two spotted carpet beetle, 147

Typhaea stercorea, 169, **170**, 171
Tyre treads, 50, 167
Tyroglyphus, 18, 20, 153, 162, **163**, 165

Uganda, 28, 112, 119
Ultra violet light, 52, 168
Undersized larvae, pupae, 34
Upton, M. S., 39
Urinals, 171
  see also Cesspits, Lavatories, Latrines, Toilets
Urine, 51, 94, 122, 125
Urogenital myiasis, 122
Urticating hairs (Dermestidae), 147
U.S.A., 15, 17, 19–20, 22, 25–7, 29, 77, 81, 90, 101, 103, 112, 122, 126, 142, 144–5, 147, 150, 156–7, 165
U.S.S.R., 94, 107
Uterus, 14
Utsumi, K., 15, 34, 156

Vacuum cleaners, 162
Valdivia (Chile), 149
Vargas, E., 12
Varley, G. C., 34, 52
Vegetable canneries, 96
Vegetation, 28–9, 41, 56–67 (case histories), 158 (overhanging)
Vehicles, 167
  see also Cars
Venom, 154, 166
Vertebrate scavengers, 17, 34, 59
Vespidae, 21, 154
Vespula, 154
Villeneuve, G., 15–6, 92
Vinagradova, E. B., 53, 105
Vincent, C., 12
Vinegar flies, 96
Vinyl, car roof, 167
Viviparity, 34, 99, 104
Voeten, R., 120
Vogler, C. H., 94
Voles, 24, 31, 73
Voluntary muscles, 14

Wales, 152
Walker, A. K., 39
Walker, T. J. Jr., 148
Wang, L.-Y., 17
Wardle, R. A., 27, 51, 104
Warehouses, 172
Wash-bowls, 96, 122
Washington (D.C., U.S.A.), 160
Wasps, 13, 20–1, 25, 53, 154
  nests, 132
Wasti, W. S., 122
Water, collecting in, 36, 41
  immersion in, 25–7, 36, 41, 50, 65, 75, 77, 96, 144
  paint surfaces resembling, 167
  proximity to, 168
  surface of, 160
  survival of ectoparasites in, 25, 162, 164, 166
Water lice, 53

Waterhouse, D. F., 51
Watrin, P., 30, 166
Webb, J. P. Jr., 166
West, L. W., 125
Wet clothes, 96, 122
Wet conditions, corpses in, 141–2
Whale, 126
Wheat stacks, 150
Wheeler, M. R., 96
Whitcombe, R. P., 117
White ants (termites), 161
White shouldered house moth, **151**, 152
Whitfield, F. G. S., 167
Whiting, P. W., 53, 103
Wigglesworth, V. B., 55
Wijesundara, D. P., 118
Willott, G., 10
Wilson, E. M., 12
Wind, effect of, 142
Windows, 33
Windscreens, 167
  see also Cars
Wings, 53, 59
  frayed in *Lucilia*, 109
  of Diptera, 70
  of insects, 52
  wagging, 85
Winter conditions, 19, 33, 73, 172
  see also Seasons
Winter gnats, 33, 73, **133**
Wolfahrtia, 99
Wolfe, L. S., 44
Wollenek, G., 11
Wong, H. R., 43
Wood, boring beetles, 146, 172
  rotten, 82, 146, 161
  urine soaked, 94
Woods, 19
Woollen goods, 147, 152
'Woolly bear' larvae, 147
Working habits, effects on corpse, 50
World War I, 51
World War II, 118
Worm holes, 79
Wounds, 52, 99, 104, 111–2, 118
  myiasis, 99, 101, 118
  therapy with maggots, 113
Wrack beds (seaweed), 81
Wrappings (around corpses, etc.), 46, 56–67 (case histories), 145
Wright, E. A., 113

Xylocopa, 154

Yovanovitch, P., 11, 56, 101

Zakharova, N. F., 53
Zimbabwe (S. Rhodesia), 117
Zimin, A., 99, 128
Zinovjeva, K. B., 53
Zoogeography, 28, 54 (map of zoogeographical regions)
Zumpt, F., 9, 80, 101–2, 104, 109, 112–3, 116–7
Zuska, J., 92

PROPERTY OF
THE WILTSHIRE CONSTABULARY
SEPTEMBER 1990